THE SILENT LISTENER

THE SILENT LISTENER

FALKLANDS 1982

D. J. THORP

SPELLMOUNT

First published 2011 by Spellmount, an imprint of
The History Press
The Mill, Brimscombe Port
Stroud, Gloucestershire, GL5 2QG
www.thehistorypress.co.uk

British Library Cataloguing in Publication Data.
A catalogue record for this book is available from the British Library.

ISBN 978 0 7524 6029 1

Typesetting and origination by The History Press
Printed in Great Britain

CONTENTS

GLOSSARY

BAOR	British Army of the Rhine
Comms & Sy Gp UK	Communications and Security Group United Kingdom
CSO	Composite Signals Organisation
dB	Decibel
Comint	Communications Intelligence
COMSEC	Communication Security
COMSO	Communication Security Officer
DF	Direction Finding
DV	Developed Vetting
Elint	Electronic Intercept
EW	Electronic Warfare
EW (Ops)	Electronic Warfare Operators
FRG	Federal Republic of Germany (West Germany)
FTX	Field Training Exercise (Troops on the ground)
GCHQ	Government Communications Head Quarters
GDR	German Democratic Republic (East Germany)
HF	High Frequency
JNCO	Junior Non Commissioned Officer
LADE	Lineas Aereas del Estado
LCU	Landing Craft Utility
LFFI	Land Forces Falkland Island
LPD	Landing Platform and Dock

LSL	Landing Ship Logistic
MOCA	Morse Operator Characteristics Analyses
MoD	Ministry of Defence
MPBW	Ministry of Public Buildings and Works
NBC	Nuclear Biological and Chemical
OC	Officer Commanding
Ops Tp	Operations Troop
PoW	Prisoner of War
PV	Positive Vetting
RAS	Replenishment at sea
RC	Roman Catholic
RDF	Radio Direction Finding
RFA	Royal Fleet Auxiliary
RSM	Regimental Sergeant Major (WO1)
RV	Rendezvous
QE2	Queen Elizabeth 2nd (ship)
SAS	Special Air Service
Sigint	Signals Intelligence
SNCO	Senior Non Commissioned Officer
STD	Special Task Detachment
SPA	Special Projects Agency
TCW	Tactical Communications Wing
Tp	Troop
WO1/2	Warrant Officer 1st Class/2nd Class
UHF	Ultra High Frequency
USSR	Union of Socialist Soviet Republic
VHF	Very High Frequency
2i/c	Second in Command

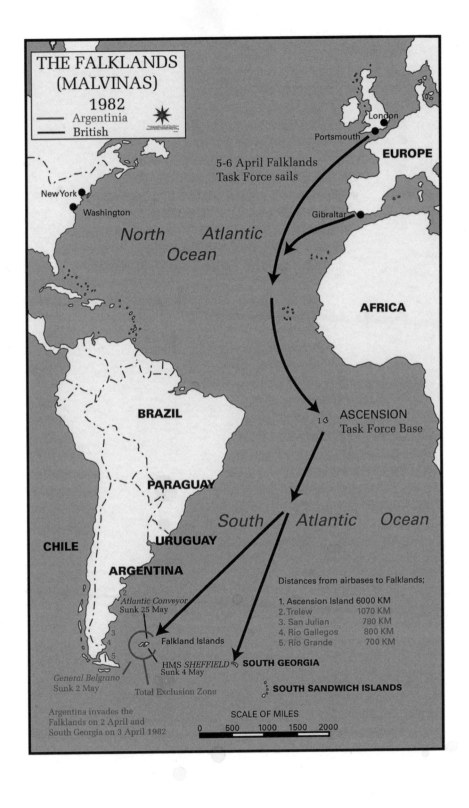

THE FALKLANDS
(MALVINAS)
1982
—— Argentinia
—— British

5-6 April Falklands
Task Force sails

London
Portsmouth
EUROPE

New York
Washington

Gibraltar

North   Atlantic
Ocean

AFRICA

BRAZIL

ASCENSION
Task Force Base
1 3

PARAGUAY

South   Atlantic   Ocean

CHILE

URUGUAY

ARGENTINA

Distances from airbases to Falklands;

Atlantic Conveyor
Sunk 25 May

2

3
4
5

Falkland Islands

1. Ascension Island 6000 KM
2. Trelew   1070 KM
3. San Julian   780 KM
4. Rio Gallegos   800 KM
5. Rio Grande   700 KM

HMS SHEFFIELD
Sunk 4 May

SOUTH GEORGIA

General Belgrano
Sunk 2 May

Total Exclusion Zone

SOUTH SANDWICH ISLANDS

Argentina invades the
Falklands on 2 April and
South Georgia on 3 April 1982

SCALE OF MILES

0   500   1000 1500 2000

PREFACE

The objective of this book is primarily to give the reader insights, alternative answers and in some cases the truth relating to certain events that took place during the Falkland Island War of 1982. In addition its content is aimed to broaden one's knowledge of a very small number of Armed Forces personnel employed in specialist duties during the period 1940–1990, who, because of the restraints placed on them by their signing of the Official Secrets Act, have seldom attracted the attention of the general public. I have deliberately liberally peppered my 'real war' memories with an overview of my military career in general during the period 1955–1988, in order to give the reader an insight into an aspect of Army life that seldom hits the headlines.

After 1990, mainly because of the wars in Iraq and Afghanistan and the more relaxed relations between the super powers after the Cold War, the role of these specialist personnel changed and through these changes, and more open reporting by the media, their existence has become more widely known. My specialist knowledge and expertise only came about after years of experience employed in the clandestine world of the electronic intercept of communication systems used by the potential enemies of Great Britain and her closest allies.

The specialists to whom I refer were actively employed 365 days a year. When on operational assignments their place of work was often hostile and on occasions very dangerous; they could be employed in the air, on the ground or at sea. Where they worked was also varied; they could be deployed to any continent in conditions, from the opulent and palatial to living and working in chicken coops, as they once did in Kenya.

Classified information pertaining to Official Intelligence Activities in all its forms and Signals Intelligence in particular, is not subject to any fixed timetables for release, such as the so-called 'Defence Notice' (D Notice) or since 1971 its replacement the 'Defence Advisory Notice' (DA Notice) and the '30-Year Rule'. The D Notice was established in 1912, bolstered by the Official Secrets Acts, to define subjects that are not cleared for public broadcast. With the progress of technology, today's DA Notices cover media broadcast content via radio, films, television and the internet, and may be applicable to other government information under the Public Records Acts. In addition, the Freedom of Information Acts do not apply to the intelligence agencies; the Acts explicitly exempt them from any obligation to provide information concerning any units of the Armed Forces 'which are for the time being required to assist the Government Communications Headquarters [GCHQ] in the exercise of its functions'.

In February 2010, the British Government presented their 'Review of the 30-Year Rule' to Parliament and the general public. Amendments to the Constitutional Reform and Governance Bill were tabled.

The reduction of the 30-Year Rule through amendments to the Constitutional Reform and Governance Bill is perhaps one of the first steps towards transparency in the field of dissemination of past highly classified information into the public arena. No sooner had the ink dried on this Bill – now an Act – when in June 2010, details of the 1946 UK–USA (pronounced ew-koo-sa) Secret Agreement was released to the British National Archives by GCHQ. This agreement, brokered with the US, led to the sharing of all signals secret intelligence between the two countries. The agreement was later extended to include the three former British dominions, Canada (1948), Australia and New Zealand (1956). The UKUSA Agreement was a follow-up to the 1943 BRUSA Agreement, a Second World War agreement on cooperation over intelligence matters – this was a secret treaty, allegedly so secret that it was kept from the Australian prime ministers until 1973 – and formalised the intelligence sharing agreement in the Atlantic Charter, signed in 1941, before the entry of the US into the conflict. While rumours suggesting the existence of such an agreement had persisted for many years, outside of GCHQ the actual document, or its content, has never been published before. The agreement itself states 'It will be contrary to this agreement to reveal its existence to any third party whatever.' With top secret codeword protection, the document was drawn up and signed by members of the forerunners of the National Security Agency (NSA) and GCHQ, the State-Army-Navy Communications Intelligence Board, known as STANCIB, and the London Signal Intelligence Board.

The Silent Listener confirms the existence and role of the Special Task Detachment during Operation *Corporate* and provides, for the first time in the history of British warfare, details of the deployment and operational role of a dedicated, but limited, ground-based electronic warfare weapons facility under the direct control of the Land Force Commander; a significant development in military history that appears to have been omitted by Professor Lawrence Freedman in his Official History of the Falkland War.

The content of *The Silent Listener* has been shown to and commented on by authorities at both Government Communication Headquarters and the Ministry of Defence; in consequence all requests from these Government Departments for changes to meet the terms of the Official Secrets Acts have been complied with. Failure to have done so may have caused damage to Great Britain's strategic and tactical military capability and aims, and risked British servicemen's and women's lives in current and future conflict.

The Constitutional Reform and Governance Act means that over the next 10 years national archiving after 20 instead of 30 years will be required by law. Though within the Act, in Freedom of Information 46 (2) it states: 'The Secretary of State may by order make transitional, transitory or saving provision in connection with the coming into force of paragraph 4 of Schedule 7 (which reduces from 30 years to 20 years the period at the end of which a record becomes a historical record for the purposes of Part 6 of the Freedom of Information Act 2000).'

INTRODUCTION

The art of intelligence gathering has been practised almost since the beginning of human societies. The need of individuals, organisations and nations to know more about others than the others know about them is insatiable. The ways and means of gathering intelligence are vast and varied, as are the reasons why groups feel the need to gather such information.

Throughout history much has been written about the world of intelligence gathering on a national level, where the target array varies from the activities of an entire nation, down to the individual on the street gossiping over the garden fence. While some people, through the press and other forms of media, like to feel they know a little about the nation's civilian intelligence gathering organisations such as MI5, MI6 and perhaps the Government Communications Headquarters, it is generally only those, because of security implications and the policy of 'need to know', who have been employed or actively involved in the various aspects of military intelligence that have any real knowledge of the role and capabilities of the Intelligence Corps and in particular that element whose specialist duties are in the collection, transcription, translation, cryptanalysis and eventual reporting of signals intelligence. In the twenty-first century it has more commonly become known as 'electronic warfare'.

The Crimea War (1853–1856), although best known for the Charge of the Light Brigade and the Thin Red Line during Battle of Balaclava on 25 October 1854, also made history as the first modern war to introduce the tactical use of the telegraph. With advances in telegraphic equipment and technology by 1901, when the Italian Guglielmo Marconi first discovered he could transmit and receive signals across the Atlantic Ocean, that department within the British Army responsible for its communications has frequently been at the forefront in the exploitation of radio transmissions. From early in the twentieth century, continual research and development within the field of electromagnetic radiation has allowed military commanders to communicate worldwide over secure means by the use of the high frequency wave bands for Morse code (less susceptible to interference) for long haul, and very high frequency wave bands for voice transmissions over shorter distances. In 1920, the Royal Corps of Signals was formed to provide communication specialists for the British Army and to this day it continues to be the Army's leader in information technology and communications.

Technology pertaining to electronic communications has undergone a radical transformation even in the last decade. The use of Morse code in the passage of radio communications is all but defunct. In a modern age of communication satellites, secure cellular telephones and the World Wide Web, radio communications via terrestrial transmitters and receivers are slowly being forsaken for state-of-the-art equipment and techniques.

1

THE TRAINING OF PROFESSIONAL COMMUNICATORS

I was born into a service family; my grandfather was a regular soldier during the First World War and eventually retired with the rank of Warrant Officer 2 (WO2) and my father retired after 30 years as an officer in the Royal Corps of Signals having served prior to, during and after the Second World War. My older brother served twelve years, including active service in Cyprus, and my younger brother served for 27 years, including several years in Northern Ireland during the Troubles. Continuing the tradition, my eldest son served nine years and completed tours of duty in Belize and with the United Nations Peace Keeping Force in Cyprus. We all served in the Royal Corps of Signals – as miners' sons went down the pits so the Thorp sons enlisted into the British Army. Upon leaving school in February 1955 at the age of fifteen I agreed to serve for three years with what was known then as the Army Apprentice School (AAS), Harrogate, North Yorkshire, followed by a further nine years with the colours in the Royal Corps of Signals.

The AAS Harrogate was established in 1947 to provide the British Army both with a small cadre of regular soldiers suitably institutionalised from a young age to fulfil the roles of future senior non-commissioned and warrant officers, and also full apprenticeship training in a variety of trades including carpenters, painters, quantity surveyors and radio mechanics for various regiments and corps. This apprenticeship training was occasionally extended to servicemen from colonial and other armed forces – during the period from the late 1940s to early 1950s the largest single contingent of foreign apprentices to serve at Harrogate was from Myanmar (Burma), with the majority training as communication mechanics (later changed to technicians). The 'apprenticeship' in most cases was of three years duration, specialised trade training culminated in the passing of City and Guilds and other specialised trade-associated examinations, alongside training in military skills and education to include General Certificate of Education 'O' and 'A' levels.

The School was renamed the Army Apprentice College, Harrogate in 1966; by this time the only corps training its 'apprentices' at Harrogate was the Royal Corps of Signals. To coincide with its change of name, the College badge was also changed to that of the Royal Corps of Signals. The final graduation parade at the College took place in August 1996.

In addition to its apprentice school, the Army also recruited 'boys', officially referred to as Junior Leaders – young men aged between 15 and 17½ years of age. While these 'boys' received full-time military training, the specialist skills relating to trade training they received was very basic. Unlike an 'apprentice' who spent three years, unless he was 'back-squadded' through illness or lack of progress before transferring, a boy entrant on reaching the age of 17½ years was automatically transferred to 'man-service'.

Having passed the apprentice written entrance exams and through a medical examination board been declared fit to serve in Her Majesty's Forces, I received instructions that I was to report to a central recruiting centre by 0900 hours on 8 February 1955, for induction into the Army, prior to travelling to the AAS in North Yorkshire.

At this time my father was a serving officer stationed at the War Office – he commuted daily by car from our family home in Chingford, Essex, to his office in Whitehall. Knowing that I would need assistance with travelling arrangements, I made the grave mistake of assuming he would take me with him at least as far as Whitehall, although secretly wishing he might drop me off at the door of the recruiting centre; after all I was only fifteen years of age and leaving home for the first time. No such luck, he informed me the evening prior to my enlistment that he would give me an early call the following morning to ensure I left in good time to catch a bus. With a small suitcase containing the minimum of personal items – I was instructed to take only items for washing, shaving and cleaning my military clothing and equipment, all clothing and other items of a personal nature would be issued on arrival – I left my parents home at the crack of dawn on a very cold and wet Tuesday morning in February to catch a Green Line bus that would take me to the Army Recruiting Office, Great Scotland Yard, where, after swearing allegiance to the Queen and receiving the Queen's Shilling, I was given a rail warrant to take me from King's Cross to Harrogate. The journey north was reminiscent of several years previously, when my older brother and I, each carrying gas masks and small cases containing most of our personal belongings, were evacuated during the Blitz to stay for several months with a family in Thornaby-on-Tees.

On arrival at Harrogate railway station, I along with other potential 'apprentices', some of whom had been waiting several hours, was met by the Duty Corporal. He checked our details against his records, confirmed no other trains were expected within the next couple of hours, then escorted us outside where our onward transport was waiting. In the pouring rain and strong winds, the small group of us were taken in an open 3-ton truck without seating to Uniacke Barracks, Penny Pot Lane, Harrogate, which was to be my home and place of learning for the next three years.

Some events and occasions, for better or worse, will always remain in people's memories and my arrival at Uniacke Barracks is one such memory. Debussing at the Apprentice School, all passengers were directed to a single-storey wooden structure known as a Spider. The Spiders – single storey construction, built from wood larch lap style panels under a gable roof, externally coated with dark brown creosote for protection against the harsh weather of the North Yorkshire Moors – both at Uniacke and across the road at Hildebrand Barracks (permanent staff accommodation) – were originally erected prior to the Second World War as temporary barrack accommodation for soldiers, with a change of use to a US field hospital during the war years.

The barracks consisted of six or eight large rooms constructed in a three or four, left and right 'herringbone' formation, with a very large central area and covered corridors linking the middle to the 'legs' of the Spider. The centre, the body of the spider, housed the ablutions, toilets, baths (no shower facilities), wash hand basins, 'Blanco rooms' and drying rooms, while the 'herring bones' or legs, were large dormitory rooms to accommodate up to eighteen occupants, nine on each side of the room. Each one of the eighteen bed spaces was furnished with a cast-iron framed bed, a single horse-hair mattress about six inches deep placed over a wire and metal support, a wooden wardrobe and a bedside cabinet.

Windows on both sides of the room were dressed with curtains in a plain khaki colour; we later discovered these curtains were for cosmetic purposes only and were never pulled for privacy. Occupants that claimed a bed space below a window were later to regret their choice because on kit inspection parades, if an individual's kit was not up to standard, it was easy for the inspecting officer to jettison the offending items – sometimes one's entire issue of military clothing and equipment – through an open window irrespective of the weather, whereas those not in close proximity to a window only had to contend with retrieving their kit from the floor around the bed.

The aisles were bare planks of wood and they were polished with brown wax polish – to maintain its shine, the floor was polished with brushes from boot cleaning kit for twenty minutes each morning by the room's occupants. The planks of wood to the left and right of centre were coloured black. I soon discovered that the colour was obtained by mixing brown wax polish with Zeebo Black Polish (normally used to polish black cast-iron fire grates); Zeebo Polish contained lead and after walking barefoot on the floor for several months we wondered why we had 'foot rot'. The floors of the surrounding corridors required regular scrubbing with soap and hot water. A punishment regular dispensed by the permanent staff (a regular soldier or National Serviceman as opposed to 'boy') was to scrub these corridors using only a toothbrush (personal property of the offender), soap and water.

No personal possession was allowed to be openly displayed and absolutely nothing was to be stuck, pinned or stapled to the walls; all notices and orders were placed in a display cabinet in the adjoining corridors. Only personal family photographs were allowed to be affixed to the inside of personal lockers. The one exception to these rules was a week before breaking up for Christmas leave, and festive decorations were permitted – there were prizes for the best decorated barrack rooms.

I previously mentioned that the barracks was once used as a US field hospital. This did not go unnoticed amongst the apprentices and this period of the camp's history was regularly the subject of many a yarn concerning the patients and staff, all based on rumour and with the inclusion and embellishment of gory details. One such story involved a US pilot who, having been shot down by enemy aircraft, was brought to Uniacke Barracks as a patient and accommodated in the same Spider as myself and other members of my term. Unfortunately, very soon after being admitted one of his legs was amputated and failing to make a full recovery he eventually died in the hospital. His legacy to the apprentices that occupied his ward was the 'Ghost of Stumpy'. Imagine some four or five years after his death, six dormitories, each with eighteen impressionable fifteen-year-olds with vivid imaginations, believing they had heard, seen or even spoken to 'Stumpy'. After rumours of such encounters it was not surprising that at night not many of them left the security of their beds to use the toilet facilities located in the centre of the Spider.

Dormitories within the Spider were allocated by alphabetical order, consequently I was in a dormitory with names starting at letter P through to Z. Having found our bed spaces, the next step was to 'march', not the word the Duty Corporal called it, to the quartermaster's stores for bedding, cutlery, and drinking mugs. Returning to our allocated bed spaces, after making up our beds and securing our personal belongings in our allocated wardrobes, we were marched to the dining hall (also known as the cook house).

My very first Army meal was London Roast, a mystery recipe consisting of minced beef, onions and other ingredients placed in a loaf tin and cooked in the oven. It was cut into slices and served with mashed potatoes, carrots and something called gravy (well it was brown and runny), followed by steamed pudding and custard. Having not eaten since breakfast that morning and the time now about 8pm, I was ravenous and ate all I was given without complaint. With the meal we had the mandatory issue of two slices of bread cut to regulation width on a hand-operated slicing machine. If stocks were running low the width of the slices was reduced. All meals came with two slices of bread and a single pat of butter, plus a single teaspoon of jam at tea. The butter and jam were insufficient to cover one slice of bread but that was the ration.

Returning to our dormitories we had a little time to become acquainted with the other occupants before lights out at 10pm. I soon discovered that I was to spend three years with a group of boys from very diverse backgrounds. Having introduced myself to two other sons of serving Army officers, coincidently one of whom I went to school with in Egypt a few years earlier, I was introduced to the boy in the bed next to mine whose name I think was Angus Taylor, from Glasgow; during our initial talk he mentioned his previous school was Doctor Guthrie's Academy. I later discovered that Doctor Guthrie's was a borstal; Angus, like several others at the school, had been up before the local magistrate's court so frequently that eventually, in an attempt to encourage him to improve his ways, he was given the choice of volunteering for the Forces or a spell in the local jail. Most given the same choice signed up.

My older brother served for twelve years in the Royal Corps of Signals, but prior to this he also spent three years at the Apprentice School, enlisting in February 1953 (Intake 53A). We were there at the same time for his last year and my first, but we had little contact; he had his circle of friends and I had mine. On the few occasions the senior boys had contact with the junior boys, it was either the junior being given personal items of a senior boy's kit to clean, or for the junior boy to be an unwilling volunteer included in the senior boys' games. These games could take the form of a race between junior boys, on their hands and knees pushing an open tin of polish with their noses along the full length of a corridor or, on the return of the senior boys to camp on a Saturday night after the pubs had closed, assembling all the junior boys on the parade ground for a regimental march pass. The Corps of Drums would be in attendance and all those on parade would be dressed in pyjamas and boots. Considering the noise of the bugles and drums, how members of the permanent staff failed to hear what was going on within the barracks at midnight was always beyond me. I believe nowadays the accepted term for this contact would be 'bullying'; we looked upon it as initiation and character building and no real harm came to any of us, but over a three-year period it was certainly a means of separating the men from the boys.

The consumption of alcohol in Uniacke Barracks was only permitted at dances and organised social functions, and then only if apprentices were over the age of eighteen. However, underage drinking outside camp was a challenge to all and considered only successful if one was able to get past the staff on duty in the Guard Room without them smelling alcohol on one's breath while 'booking' back into camp. On the other hand, smoking was permitted on reaching the age of sixteen. To legally smoke – but only outside of buildings and in designated areas – written permission had to be obtained from parents or guardians. On receipt of written permission a 'Smoking Pass' would be issued; the pass had to be produced if requested by a member of the permanent staff or an apprentice NCO. During my early period as an apprentice, the Regimental Sergeant Major (RSM) was Stan Longsborough of the Coldstream Guards. The RSM had a dog that regularly accompanied him to work, and when it needed exercise he would walk it through the barracks where he would frequently find apprentices smoking outside their barrack accommodation. The RSM would challenge the smokers to produce their pass, but to my knowledge he never caught an apprentice without one for the simple reason that he took the same route. An apprentice without a pass had plenty of time to remove any evidence that could incriminate him.

As apprentices, all our activities were strictly timetabled. I was down for physical training during the last period of the day on Thursdays. Nothing extraordinary about that except that Thursday was, in those days, special because it was pay day and the only weekday that those apprentices with less than two years' service were allowed out of barracks. Once the work day was over it was a mad scramble to get changed from uniform into blazer and flannels, rush down to the guard room where we were inspected by the Guard Commander who checked that our dress conformed to civilian dress regulations before allowing us to book out of barracks. If we moved quickly enough, we would be able to catch the 6pm bus to Harrogate in time to make the second showing of the film at the local Odeon cinema. The next bus would not be along until half an hour later and would have meant missing the start of the film. Waiting for a later showing was out of the question because we had be back in barracks by 10.30pm.

Our intake seldom made the 6pm departure because our instructor, Corporal Johnson of the Army Physical Training Corps, always took delight in finding fault with our efforts and delayed ending the class for 30 minutes or so. After several weeks of being kept behind after the class should have finished, the intake, all thirteen of us, decided that enough was enough and the following Thursday we would not attend the class. Instead we returned to our barrack rooms and lay on our beds until it was time to change out of our uniforms to go into town. But when we attempted to book out we were taking into 'close arrest' and locked up in the guard room. We were initially held on the grounds of being absent from our place of duty. Apparently when we failed to turn up at the gymnasium, the guard was alerted and searched all the possible places that we may have been hiding, everywhere except our barrack rooms otherwise they would have easily found us. Later that evening we were released from 'close arrest' and placed on 'open arrest'; this entailed reporting to the guard room first thing in the morning, at lunch time, after tea and last thing at night. After several days, including the weekend, we were charged and appeared before our Company Commander.

The Company Commander was Major Shenton, Royal Artillery and the Company Sergeant Major was CSM (Harryboy) Harrison, Coldstream Guards. Thinking that because we had spent several days on 'open arrest' this would be sufficient punishment for the crime, none of us were unduly worried. Eventually CSM Harrison marched us into Major Shenton's office who then asked the CSM to read out the charge. After reading all thirteen personal details he went on to say we had been charged under a section of Queen's Rules and Regulations for wilfully failing to attend a parade and under this particular act were accused of mutiny, and if found guilty, the offence would be punishable by death. On the word 'mutiny' I swear all that could be heard were thirteen pairs of knees knocking together. All of us were then asked if we had anything to say; all we could think about was the firing squad. Eventually we were marched out of the office, to return one at a time. After each boy had been seen by the OC he left the office by a different route so he could not relate the nature of his interrogation to the other boys.

I was the last to return for a second appearance before the Company Commander and was accused of being the ringleader and encouraging the others not to attend the class. This was not true; the idea of not attending class had been someone else's idea but I was not willing to say who it was. Having been warned about my behaviour and there not being the slightest hope of a second chance in the future, I was presented with the chevrons of a lance corporal and informed that as of now I was being promoted with the proviso that I lead my subordinates in the direction the Army would expect me to, not in a direction of my own choosing.

Promotion for an apprentice in the second term was at the time unheard of and since my retirement from the Army in 1988, I have often wondered what my life would have been like if I had been dealt with like the other twelve 'mutineers'. Would I have stayed in the Army for more than 30 years?

On commencement of my apprenticeship in February 1955 (Intake 55A), I was initially selected for training as a radio mechanic. During our first six weeks of basic training we were given an insight into our chosen trades. From this brief trade introduction I learnt that for the next two-and-a-half years most of my working day would be spent in a classroom, studying radio theory and mathematics. However, when I saw what the role of a communicator had to offer, my apprenticeship changed to that of an Operator Wireless and Line. A qualified Operator was able to transmit and receive Morse code, touch-type for keyboard work, read Five-Unit Murray code and a host of other practical skills coupled with theoretical knowledge associated with radio communications.

About one year into my trade training the War Office (later the Ministry of Defence) decided to no longer train its Royal Corps of Signals personnel as Operators Wireless and Line and introduced the new trade of Radio Telegraphist – this meant combining the ability to receive Morse code with touch-typing on a typewriter (much later on a computer), as opposed to copying by hand, thereby increasing the speed of receiving Morse code from around 12 words per minute (one word consisted of 5 characters, therefore 12 words per minute equalled 60 characters) as an Operator Wireless and Line, to receiving and transmitting Morse code letters at more than 30 words per minute and Morse code figures in excess of 35 words per minute. The higher speed for figures was because there were only 10 figures compared to 26 letters and figures were easier to identify particularly when transmitted as coded blocks of 5 figures. Operational speeds in excess of 30 and 35 words per minute are possible but only machine-aided. The simplest way to increase Morse speed is to record the message text on a tape recorder, then transmit at a much faster speed; conversely, record at the faster speed and play back at reduced speed. A small number of apprentices that failed competency in high-speed Morse code transmission were later trained in additional skills and became Special Operators.

The introduction of the new trade of Radio Telegraphist progressed so fast that the administrators for this new trade, both service and civilian, were unable to proceed at the same pace; consequently on completion of our apprenticeship there was no civilian qualification available or applicable to the standards reached. As one of the first six apprentices to qualify as Radio Telegraphist, I failed to achieve any form of civilian recognition or qualification, which after all was one of the main reasons I volunteered to enlist as an apprentice in the first place. However, for 'trade pay' purposes, Radio Telegraphist was classified as an A Class trade, while Operator Wireless and Line was a B Class trade; this meant that the six of us were financially better off at the end of our apprenticeship compared to those that had qualified as operators before us.

2

APPRENTICE TO VETERAN

As a qualified Radio Telegraphist my first posting on leaving Harrogate in February 1958 was to a signals regiment in support of 2 Signals Division where my older brother was also serving. It was initially located in the Düsseldorf area of West Germany, but later, on amalgamation with another division, my unit relocated to the small town of Bunde in central West Germany. Life at Bunde appeared to revolve around the consolidation of our specialist trades in preparation for a possible war against Russian and other communist forces; this was fine when on deployment outside barracks, but when in barracks, there was only the cleaning and maintenance of vehicles and equipment to be carried in preparation for the next exercise. Life proved to be extremely quiet and less hectic compared to that of my first three years spent in a military training unit. Realising after only six months that as a serving communicator with the British Army on the Rhine (BAOR) there was more to life than military manoeuvres and cleaning vehicles, I submitted an application to attend an Arabic language training course that I had seen advertised on the Squadron notice board. After two interviews by officers of the Army Education Corps, I was accepted for training on the next course.

My choice of Arabic as a second language was in part due to the fact that as a child, during the period 1948–1950, I lived with my parents in Egypt where I picked up a smattering of the language. At the time of applying for the language course I was aware that previous courses had been held at the Berlitz School of Oriental Languages, London, with some students going on to its sister school in Beirut, Lebanon, to gain practical language experience. However, changes had occurred between the times of submitting my application to its approval, which resulted in my course being held in London, not at the Berlitz, but in a room on the fourth floor of Kings Buildings, Dean Stanley Street, accommodation owned by the War Office. The language teacher for four hours' tuition Monday to Friday was a lovely woman of Middle Eastern origin known as Sit Ray, who was married to a serving British Naval officer. Sit Ray had previously been employed with the Berlitz School as a language tutor for Army students.

For the six months of the course our accommodation was with the Coldstream Guards in their barracks at Bird Cage Walk, in a room to the rear of the barracks used as accommodation for those service personnel in transit. Our personal possessions were frequently stolen because we had no secure storage space, and it was terribly cold during the winter because the room had no form of heating and the one and only window had most of the glass broken or missing. Our dress as students was smart civilian clothes for which we were paid an allowance of one shilling a week, and we were issued with a pass that allowed us to use the rear entrance and exit of the barracks along with bandsmen, other civilian staff and those considered not suitable to pass through the main gates.

Those responsible for getting the course up and running and ensuring that the accommodation used for our formal four-hour daily tuition with Sit Ray was up to the standard expected of a classroom, in their haste to get the course started, failed to provide adequate accommodation for 'home study' – homework. With a minimum of four hours homework Monday to Friday, and more at weekends, the atmosphere and facilities in our allocated barrack accommodation, as previously described, were far from ideal. However, we did find that the Union Jack Club provided warm and comfortable surroundings and this is where we spent most of our 'home study' time.

On completion of this language course I was to discover that once again I had completed a course without any form of civilian recognition. It was apparent that the only civilian Arabic language examination open to us on completion of the course was an obscure paper set at GCE 'O' Level with all questions relating to the Koran, or another similar paper specifically for students whose mother tongue was Arabic and learning English as a second language. As the five of us students were neither studying the Koran or enjoyed Arabic as a first language we had to be satisfied that our ability to read an Arabic language newspaper was sufficient. In the War Office's defence they did find an acceptable college willing to set an exam based on the content of our specific studies, but six months of preparation time was required by this college, time which we did not have. I believe that later courses benefited from the many administrative mistakes made on mine.

Prior to commencement of the language course, it was never explained that on successful completion all non Intelligence Corps personnel would be transferred to the Intelligence Corps prior to commencement of special duties. When I discovered that to be transferred would result in a reduction of pay – the trade of linguist was a 'B' Class trade for pay banding, while my current Royal Signals trade was an 'A' Class – with marriage looming, I could not afford to transfer to the Intelligence Corps and had to decline the offer of a change of corps. As a result of my actions I became the first language-trained soldier to pass through the Intelligence Corps depot at Maresfield, East Sussex, without changing my cap badge.

On completion of the language course I married Margaret during my pre-posting embarkation leave. After this short spell of leave, with three other Army students from the course (the fifth student was a civilian employed as a Russian instructor at a Service language school) I found myself in Cyprus.

The flight to Cyprus was by the Transport Command, the RAF's own de Havilland Comet 4 aircraft, the first military jet aircraft taken into service for air trooping and, in comparison to other trooping aircraft, a much more luxurious and faster means of travel. On arrival at Nicosia Airport, the four of us were transferred to the Forces transit camp located at Wayne's Keep, just outside the town of Nicosia. Wayne's Keep was a tented camp with very limited facilities, and it introduced us to the 'Char Whallahs' – men of Asian origin, mainly Indian or Pakistani, who, after the Second World War, 'followed the Flag' throughout the Middle East serving tea, known locally as 'gunfire', and curry to British servicemen from about 5am to 10pm daily.

Our transit camp was only some 40 miles from Four Mile Point, Ayios Nikolaos, the location of 2 Wireless Regiment (later changed to 9 Signal Regiment). However, it took four days to get us there because transport was not available. On arrival we spent the next three weeks doing nothing because our personnel vetting process had not been completed prior to leaving the UK.

On the completion of a special vetting processes – then known as Positive Vetting (PV) later changed to Developed Vetting (DV) – and the signing of various documents to ensure that within my personal life and family background there were no hidden skeletons that suggested I might contravene the Official Secrets Acts, I spent the next three years on specialist technical duties that the MOD has indicated it does not want defined..

At the time of our arrival in Cyprus, apart from restrictions on entering towns and curfews, and with the Eoka troubles almost over, the island was preparing itself for independence from the UK. After this was granted on 16 August 1960, Cyprus proved to be a beautiful and idyllic place in which to live. Access within and around the island was open and without restriction of movement, and the pace of life was slow. Everyone appeared to be enjoy themselves; it had its comic moments, such as when Archbishop Makarios (at the time the President-elect) visiting a detached location of my parent unit, and while attempting to hammer his crook into the rocky ground and getting nowhere fast, declared to all those present that he needed this particular site because it had such lush and fertile soil.

During this tour of duty, having been promoted to the rank of Corporal, I eventually transferred from the Royal Corps of Signals to the Intelligence Corps in the hope of improving my career prospects. At this early stage of my military career I was in a Catch 22 position – my parent corps the Royal Corps of Signals had very little hold over me, therefore promotion prospects were extremely limited and, because I was from another corps, the Intelligence Corps could do very little for me.

Therefore, being keen to advance, I had little choice but to apply for a change of cap badge and in June 1960, I was re-badged Intelligence Corps with the rank of Private, with immediate promotion to the rank of Lance Corporal. Within three months I was back to my original rank of Corporal. In hindsight, it was one of the smarter moves I made during my service career. In comparison to other corps, promotion through the ranks of the Intelligence Corps was quick. Several of my colleagues who had been called up for National Service in the Intelligence Corps had made sergeant in just two years.

In June 1962, on completion of my tour of duty in Cyprus, I was posted to a sister regiment, 13 Signal Regiment located at Birgelen, a small village about 200 yards inside West Germany's border with Holland and between the villages of Rothenbach (the border crossing point) and just outside of the small town of Wassenburg. 13 Signal Regiment can trace its origins back to 1934, when it started life in Aldershot as 4 Wireless Company. From here the unit had many changes of unit title and location of barracks before being renamed No 1 Wireless Regiment stationed at Peterborough Barracks, Gluckstadt, near Hamburg, in 1947. The Regiment moved to its permanent location at Mercury Barracks, Birgelen, in May 1955. It was designated 13 Signal Regiment (Radio) in September 1959, and remained at Mercury Barracks until its disbandment in November 1989.

The barracks that housed this specialist strategic unit was unique in so much as it was the only known British Army barracks not situated in the Middle and Far East built with verandas and storm ditches. Having decided on the site for the barracks, the next thing was for the Ministry of Public Buildings and Works (MPBW) to produce architects plans. The story originating from the first occupants of the barracks was that taking the easy way out, apparently staff at the MPBW found in their archives architectural plans of a barracks to house approximately the same number of troops as the specialist strategic unit; consequently the barracks were built using these plans. Unfortunately, the plans were for a barracks in the Far East.

Stationed in Germany where knowledge of the Arab language was of no use to me or those with whom I worked, I retrained on the job and eventually became an expert in the field of Radio Direction Finding (RDF). In addition to a knowledge of and experience in the field of signal communications, it was also essential to have a good grounding of mathematics and experience in the practical use of slide rules (the forerunner of the pocket calculator) in order to work out standard deviation, systematic errors and angles.

On completion of my tour of duty with this regiment, I remained in West Germany and spent the next two years with 225 Signal Squadron, an independent squadron located in the mountains above the small West German town of Schafholdendorf (south of the town of Hamelin of Pied Piper fame). 225 Signal Squadron was formed in 1958 as 1 British (BR) Corps Mobile EW Squadron, co-located with the then 1 Wireless Regiment at Mercury Barracks, Birgelen.

In June 1964, the Squadron relocated to Schafholdendorf; I joined a couple months later. The barracks in which the unit was housed had been a Second World War Luftwaffe glider training unit. Throughout the war the existence and function of this unit was never known by the Allies and it was not until after the war that it was discovered by accident by a couple of US servicemen who, when out for a drive, observed some eleven miles away on top of a small mountain what they thought was a monastery. They decided to take a closer look and discovered a barracks that had been abandoned in a hurry, leaving the place almost completely intact and with several gliders on launch pads awaiting jettisoning into the valley below the mountains.

The barracks had been previously occupied by a specialist RAF communications unit before being transferred to the Army. The main role of my new squadron was communications security (Comsec), the policing of radio communications by monitoring the communication networks within the 1st BR Corps for obvious breaches of security when elements of the Corps were deployed on field exercises. On occasions, during brigade and corps level exercises, the Squadron would assume the role of an enemy signals unit in the 'jamming' and 'intrusion' of 1st BR Corps exercise communication networks.

Unfortunately the 'jamming' and 'intrusion' roles were seldom practised because those responsible at 1st BR Corps for the planning and directing of exercises, particularly those that involved troops on the ground, frequently complained that my unit found it too easy to cause havoc with their exercise scenarios. One such instance was during a particular exercise scenario when tank troops had taken up their positions to advance towards their objectives, unknown to my unit.

The advance stage was to be a paper exercise only while in reality the tanks were to be placed on low loaders and transported to the next stage of the exercise by road. Not being aware of this, after successfully breaking into and taking control of the command network of the tanks communication network, an instruction from my unit for the tanks to advance towards their objective was given, which the tank commanders duly carried out. This successful 'intrusion' caused thousands of German Marks worth of damage. After this particular disruption, all future Corps exercise participants were notified in advance when my unit was participating; subsequently the mention of our participation greatly improved the overall security of 1st BR Corps communication networks.

When not participating in training exercises, this unit had a small Sigint role against the communication systems of the East German Armed Forces. In 1967, the Squadron relocated to Langeleben, west of Braunschweig, where it remained until amalgamating with other units on the formation of 14 Signal Regiment (EW) in July 1977.

Having served overseas continuously for seven years, I was posted back to the UK where my specialist duties were primarily in the field of analysing and reporting of communications of Russian military networks in the USSR. This tour of duty proved to be quite sedentary allowing me to recharge my batteries after the hustle and bustle of regular manoeuvres and Army life in general while serving with BAOR.

In August 1968, Czechoslovakia was invaded by members of the Warsaw Pact countries. The department in which I was working at the time of the invasion had more than a passing interest in the invasion but my interest was to be very short-lived because I and my family were once more on the move.

Two years in England then it was back to 9 Signal Regiment in Cyprus again where, because my Arabic language expertise was a little rusty and because of my past training and experience in the field of signals communications, I became involved in a particular form of intelligence analysis that once again I cannot describe.

Life in Cyprus during my second tour of duty had changed considerably from that of the immediate post-Eoka period. There had been a period of tension between the Greek Cypriots and the Turkish Cypriots, the United Nations peacekeeping troops were stationed on the island in large numbers and for good or bad, the island had stepped into the twentieth century in the areas of social, economic and commercial advancement. Once again I was fortunate to serve during a period of relative peace and stability and freedom of movement between the Cypriot communities. I left long before the major conflict in July 1974, which resulted in the partition of Cyprus.

In 1970, I was fortunate to be selected through the NATO Exchange Forces programme to serve with the 1st Canadian Signal Regiment, Canadian Armed Forces in Kingston, Ontario. My tour of duty in Canada happened to coincide with the Canadian Government's problems with Le Front de Libération du Québec, or the 'FLQ' as it was more commonly referred to.

Prior to my secondment to the Canadian Armed Forces, the British and Canadians had for many years participated in an exchange program, whereby personnel from both countries were cross-posted to complete a full tour of training within their specialist fields. The Canadians appeared to take up this opportunity more than the British; while I was the only member of the British Intelligence Corps on exchange in Canada, the

Canadian Intelligence Corps had three of its members serving with British units in England and West Germany.

In the late 1960s the Canadian Government decided the country no longer needed a separate Army, Navy and Air Force and combined all three arms to create the Canadian Armed Forces. This armed force was charged with the defence of the homeland against external or internal aggressors; consequently in April 1971, when the inmates of Kingston Penitentiary rioted, it was they rather than the Province of Ontario Police or the Royal Mounted Police who deployed to provide an armed guard encircling the main building in which the prisoners had locked themselves. I was not allowed to form part of this armed ring of troops and had to remain back in barracks assisting with logistical support. After four days the riot ended when troops landed on the roof of the main building from helicopters and forced an entry, by which time the rioters had killed two of their fellow inmates.

The exchange program between the British and Canadians lapsed in the late 1970s, when the Canadian Armed Forces adopted a closer training arrangement with their opposite numbers in the US Armed Forces. Perhaps the highlight of this exchange tour of duty was the three weeks on detached duty to Alert, Ellesmere Island, then the most northerly inhabited place in the world. During my visit I did not see daylight from the time I arrived to the time I departed.

Returning to England in 1972, I served with 223 Signals Squadron in Flowerdown, Winchester, where I was operationally employed on the analyses of intercepted radio communications of ground forces from countries within the Warsaw Pact. 223 Signal Squadron had one of the shortest reigns as an independent specialist signal squadron. Reformed in April 1967 primarily to meet the requirement of a 'home posting' for those specialist tradesmen with personal problems, after only ten years the Squadron was absorbed by 224 Signal Squadron, a specialist trade training establishment located at Woodhouse, Loughborough. My tour of duty in Winchester was cut short in 1974 on promotion to the rank of Warrant Officer 1st Class (WO1) and selection as Senior Instructor at 4 Communications (Coms) Unit in Woodhouse, Loughborough.

Having spent most of my married life outside of the UK, now at the age of 34, promoted to the substantive rank of WO1 and with a further six years to serve to complete 22 years regular service, the first opportunity to retire on pension, I thought that I was now in a position to continue at a slower pace, having reached what I thought at the time to be the pinnacle of my Army career. I had proved the recruiters correct in their assumption that the aim of the Army Apprentice School was to produce highly competent and professional tradesmen as well as the future warrant officers and senior non commissioned officers for the British Army.

3

MY CAREER AS AN ACORN

In military ground communications, for security reasons, it is not normal to reveal the ranks or names of those in conversation with each other. Therefore, in order to protect the anonymity of the speakers, a series of cover names are used: the unit commander 'Sunray', artillery commander 'Sheldrake', intelligence officer 'Acorn'.

On arrival at a new unit, it is customary to be interviewed by the commanding officer or his recognised deputy. Therefore, as I arrived at 4 Coms Unit on a Sunday evening, I was told to prepare for interview first thing the following morning. Welcoming me to his unit, my new CO informed me that I was required to report to the MoD the following day for an interview with the Brigadier responsible for DI24(A), the sponsor branch for my specialist environment. On Tuesday morning I found myself in the Brigadier's office, drinking coffee with him. The Brigadier commenced the interview by informing me that I had arrived at a critical point in my career – on being told this by a very senior officer, I honestly thought I was in trouble – far from it, because he then informed me, subject to my acceptance, I was to be commissioned into the Intelligence Corps. Of all the possibilities running through my mind, being offered a commission was not one of them, particularly as I had only 24 hours beforehand started a tour of duty at a new station. On being given this news, I naturally had a few questions but I was told by the Brigadier to remain quiet, and in ten minutes respond to the question 'would I accept a commission'. When finally asked the question, naturally I replied 'Yes'.

The offer of a commission did come with strings attached. Firstly I had to attend a six-month Russian language course; secondly, at the end of the course I would be commissioned and immediately posted to BAOR for a new tour of duty. The language course was held in Cheltenham, under the auspices of the Army who provided the tutors both civilian and military, with students accommodated in a local hotel. Unfortunately, half way through the course I became seriously ill, was admitted to hospital and in all missed over a month of the course. Thinking that I would be back-coursed to catch up on the work missed, I was surprised when DI24(A) stressed that I should not be unduly worried about completing the course but my priority was to get fit in preparation to be posted, after commissioning, by the pre-arranged date in June 1975.

Commissioning from Warrant Officer within the Intelligence Corps was very straightforward. In other corps and regiments, to allow for a smooth transition from one 'mess' to another, it was normal for the individual to be commissioned to attend a two week 'knife and fork' course on etiquette and how to behave as an 'officer and a gentleman'. On 16 June 1975 all I had to do was report to the depot of the Intelligence Corps, which in those days was in Ashford, Kent, in the parade dress of a Lieutenant, and painlessly be administratively discharged from the ranks prior to being commissioned as a Lieutenant in the Intelligence Corps. On completion of the paperwork I was taken to the Officers' Mess and officially 'dined in'.

In the third phase of my military career I joined 226 Signal Squadron, another small independent Signal Squadron, this time located at Wesendorf, West Germany, within a few kilometres of the East/West German border. The site was originally a Luftwaffe airfield from where, during the Second World War, German bombers were loaded with bombs prior to setting off on raids against British targets. Alongside the airfield was a rail station which allowed the bombs to be delivered straight from the armaments factories to the airfield where they were stored, some in underground bunkers. After the war part of the airfield was handed over for use by the British and eventually occupied by a small RAF unit. This unit was one of several similar sites located in close proximity to the border to carry out electronic intercept (Elint) as well as forming part of a base line for the location of targeted electronic transmissions by direction-finding techniques of Warsaw Pact ground-based surface-to-air missiles units, both strategic or deployed in a mobile role.

The site, now transferred to the Army and occupied by 226 Signal Squadron, had a strategic radar operational centre housed in the old railway station from where it carried out its operational role of maintaining continuity on, and monitoring the operational capability of, Soviet Surface to Air Missile Systems. The variety of radio antennas available for use in the intercept of targets were housed in a tall metal-framed construction covered in plastic sheeting known as a Martello Tower. The local inhabitants from the surrounding area were convinced that the tower was really a silo and housed a nuclear rocket that in the event of hostilities would be fired towards Russia; it virtually eliminated the need for the Squadron to have a cover story explaining why they were located in that particular part of West Germany.

Not long after the Squadron took occupation of the site, it became apparent that in the event of any serious hostilities between East and West this strategic site would be of no useful purpose. Due to its closeness to the forward edge of the battle area, its operational life could be counted in hours at most, because if not reduced to a heap of rubble from aerial bombardment, advancing troops from across the border would quickly render the facility useless with a ground attack.

The idea of a missile-locating capability in the armoury of 1st British Corps appealed to the Ministry of Defence but unfortunately the vulnerable location of this asset did not. As the Operations Officer, I spent two years helping to transform this vulnerably located strategic unit, with its Second World War vintage, highly classified and sensitive equipment designed to be operational from a permanent static location, into a mobile tactical unit to be moved around a battlefield as and when required.

In order to justify its existence, from its limited resources the Squadron continued operations from its strategic location keeping a watching brief on enemy targets located within East Germany, as well as mobile tactical operations to become proficient in its primary role – identifying and locating enemy radar installations. In order to practise its mobile role, the Squadron's operational element often exercised against artillery assets belonging to 1st British Corps. With the introduction of the Rapier Missile System into the Corps' area of responsibility, the Squadron was particular busy in proving, or disproving, that this new weapons system could be vulnerable to a potential enemy and advising the Royal Artillery on changes they should make to their standard operational procedures.

In 1979, the tactical Elint squadron located in Wesendorf and the Sigint/Comsec squadron I served with in Schafholdendorf, which by this time had been relocated to Langeleben (also within close proximity of the East/

West German border) amalgamated to form 14 Signal Regiment (EW). 14 Signal Regiment was formed in 1959, with responsibility for worldwide communications on behalf of the British Army. The role of providing long haul communications was later transferred to the RAF, leaving 14 Signal Regiment surplus to requirements until 1977, when it was reformed as 14 Signal Regiment (EW) with the mission to provide Commander 1st British Corps with electronic warfare support. By 1978, the new regiment had established its headquarters in Scheuen just outside of the old town of Celle. Since its formation the regiment has frequently changed its headquarters, locations and missions. Initially it provided EW support to 1st British Corps, in 1992 its role was to support NATO and the British elements under its command, now the Regiment is the only British Army regiment capable of conducting sustainable EW in support of national operations worldwide.

The barracks at Scheuen had a macabre history – a railway line terminated at the rear of the barracks which had previously been used to transport Jews from Eastern Germany and Poland to the Belsen concentration camp only some fifteen miles away; the last part of the journey was made by road. I first visited Belsen in 1958, when I and another radio telegraphist were temporarily seconded to an element of the German Army located in the Hohne tank training area, to train the German Army range operators in the use of British radio procedures in preparation for when they were to assume control of the ranges. This was only some thirteen years after the camp's discovery at the end of the Second World War. With its mounds of earth covered with heather and a small sign showing the number in thousands buried below, it took on a strange aura of somewhere set apart that commanded attention. In those days silence surrounded the total area in and around the camp, people spoke in whispers, nothing appeared to move and even birds were conspicuous by their absence.

On the amalgamation of this new regiment my previous appointment as Operations Officer of the Elint element within the new regiment was changed to Senior Intelligence Officer, with responsibility to ensure that target tasking and the Regiment's specialist operational role was strictly carried out in accordance with instructions issued by Corps Headquarters and national centres. My secondary duty was as Communications Security Officer (Comso). Due to the sensitivity, the extremely high security classification and need-to-know policies placed on the raw, semi-processed and reported material, every serviceman and woman had to have a current personal security clearance, unique to the security level determined by the security classification of the material to be handled. The responsibility for ensuring an individual's clearance was of the correct level for the job in hand lay with the Comso. The Comso was the only person within a Sigint regiment or squadron allowed access to individual's personal security clearance data, and as such was responsible for ensuring clearances had been received or despatched, prior to posting in or out of individuals. Another of the Senior Intelligence Officer's secondary duties was, as an attached member staff of the Corps Intelligence Cell, to provide simulated intelligence reports on, and play the role of, a potential Warsaw Pact enemy during 1st British Corps field tactical exercises.

After four years of mobile tactical training, in September 1979 I was posted as the Second in Command (2i/c) and Operations Officer of a small sub unit located in West Berlin, where I was responsible for the reporting and analyses of mainly ground-based tactical voice communications of various Warsaw Pact countries. The unit was located in Teufelsberg (German for 'Devil's Mountain'), an artificial hill at a height of 80 metres above the surrounding Brandenburg plain and the highest spot in Berlin. The hill was built by the Allies from the rubble cleared from the city's buildings that had been destroyed during the Second World War.

This squadron, parented by 13 Signal Regiment located in Birgelen near the Dutch/German border, originally started its operational role as a detachment, Royal Signals Detachment Berlin. The detachment later moved to accommodation housed on the operational site at Teufelsberg. Although Teufelsberg Hill was located in the British sector of Berlin, the buildings on it were built by the US for use by their National Security Agency with space sub-let to the RAF, who in turn sub-let an area of the basement to the British Army. Prior to the buildings being erected on the site, the hill was used for skiing by the residents of Berlin. All operations at Teufelsberg ceased after the fall of the Berlin Wall and the collapse of the Soviet Union in 1991. At the time of writing the accommodation on Teufelsberg was still standing but in a very dilapidated state. If only walls could talk, what intriguing and wonderful stories these could tell!

Living in Berlin prior to the German reunification and the destruction of the Berlin Wall was an unforgettable experience. West Berlin overtly displayed signs of opulence, a city rich in history, well advanced into the modern technological age. For an officer of the British Army, serving as a member of an occupied force had much in common with the lifestyle of those officers on secondment to the Indian Army prior to the start of the

Second World War. Living the life of the pseudo British Raj was a great social whirl with frequent visits by British royalty, MPs and other foreign dignitaries. Visits were frequently in conjunction with ceremonial parades within the Olympic Stadium, Brandenburg Gate and Spandau Prison where Rudolf Hess was incarcerated from the end of the Second World War until his death in 1987. Having lived in or visited other large towns and cities around the world nothing compares to the atmosphere and sinister air of intrigue felt by living behind the Iron Curtain in a walled city during the Cold War. Since leaving, I have often recommended that anyone preferring city life should live in Berlin, which lives 24 hours a day.

As mentioned above, members of the British Royal Family were fairly frequent visitors to the British sector of Berlin. One such visit to my unit was by Her Royal Highness the Princess Anne, the Princess Royal, in her official appointment as Colonel-in-Chief, Royal Corps of Signals (her father Prince Philip, Duke of Edinburgh is Colonel-in-Chief of the Intelligence Corps). Approximately one week prior to her visit, one of her personal protection police officers visited our basement accommodation to check the security of the route to be taken, for which I was his guide. On entering our secure area, which incidentally was fully carpeted, we were confronted by a lake of dirty, smelly water caused by flooding from the sewers. Wading through the foul water we completed our walkthrough, and as we parted I told him not to worry as 'It would be all right on the night.'

A few days later the royal visitor arrived at our unit. As the Operations Officer, I was given the task of explaining the role of the unit to Princess Anne. On completion of my briefing I made a horrendous faux pas by asking quite politely (and in accordance with my past training as an instructor), if the Princess Anne had any questions? An immediate response came from the General Berlin Brigade, who was hosting the Princess, to the effect that I, as a junior officer, should not ask royalty if they have any questions. Princess Anne's response to the General's comments was 'Oh yes he can, and I have plenty of questions I would like him to answer.' At the time of her departure Princess Anne thanked me and before turning away said 'Yes, it was "all right" on the night.'

In December 1982 I left Berlin, returning to the UK to take up the appointment of Squadron Second in Command and Operations Officer of a predominately Royal Corps of Signals operator training squadron at Communication and Security Group UK (Comms and Sy Gp UK), Loughborough. On my return to this Squadron after active service in the

Falkland Islands, I became the first such officer to command a squadron of Royal Signals' personnel employed in the training of special telegraphists.

As much of the rest of this book deals with my time in Falklands, I shall not go into detail here, but rather give a quick description of my career afterwards. In 1983 I was transferred from Comms and Sy Gp UK to one of its detached squadrons in Cheltenham, Gloucestershire. I was originally posted for training and familiarisation on targets within the operational sphere of responsibility of a unit I was to rejoin in 1984. My short tour of duty was punctuated by taking time away from this familiarisation training firstly to investigate the sinking of the Argentinian ARA *General Belgrano* during the Falklands War, and secondly to cover, on behalf of National Centres, Operation *Urgent Fury*, the Invasion of Grenada as ordered by President Ronald Reagan.

The Invasion of Grenada commenced at 0500 hours on 25 October 1983, with some 7,000 US troops plus reinforcements made up of troops from the Organisation of Eastern Caribbean States. The land battle lasted several days resulting in the deaths of 19 and the wounding of 116 US and coalition forces, and the deaths of 25, with 59 wounded of the opposing forces. This invasion, the first major operation conducted by the US military since the Vietnam War, was fairly short in duration and ended in December 1983. My contribution was to act as a link with interested parties in the UK.

In 1984, I returned for the last time to West Germany and served as the Senior Intelligence Officer at 13 Signal Regiment, Birgelen, responsible for the Regiment's role in the tasking, collection, analyses and reporting of ground-based radio communications emanating from the USSR.

Family commitments, in particular stability of my children's future education and aging parents, dictated I serve this tour of duty unaccompanied. This time Dad went off to 'boarding school' while the children got to live in the family home. This enabled two of my three children, who had spent a considerable amount of time being educated in boarding schools, time to place their roots somewhere they could eventually call home, rather than a place they visited during the holidays.

I served my final tour of duty in England, analysing and reporting on the communications of the *Glavnoye Razvedyvatelnoye Upravleniye*, better known as the GRU or the department of the Russian Ground Forces responsible for Soviet Military Intelligence. In reality, not as exciting as a John le Carré novel. This final tour of duty allowed me time to unwind, look at life from a different perspective and familiarise myself with civilian life.

Looking into the future, career-wise, what did I have to look to forward to? Not a lot. I had served alongside veterans from the Second World War and National Servicemen, an experience not to have been missed. I had served with foreign armies and my career had taken me to places that I would not ordinarily have visited. Within my specialist field I had served with all units employed in the strategic or tactical roles of the specialisation. I went to war and the end results our small team under my command achieved were significant; on reflection this was the most rewarding period of my career.

While I believe that I am flexible and versatile enough to accept change, I also believe that it should only be for the better. One area of progress that I had serious doubts about was the Army's selection of its future senior officers. For most of my career I was commanded by senior officers who were, in the main, selected from the 'upper classes', almost invariably products of the landed gentry and a public school education. These types of officers, because of the way they had been brought up, were groomed from an early age for service life, trained in the arts of leadership, command, control and discipline. In the 1970s and into the 1980s, selection of potential officers changed and the senior officers of the future were recruited from the grammar schools, with little or no knowledge of what was expected from a senior officer. Not all ex-grammar school officers failed to meet the highest expectations, but if only two or three did, then it was two or three too many. Because of their schooling in the public sector they lacked experience of living an 'institutional' way of life. Consequently they lacked the maturity to treat soldiers as adults and their style of command was on a par with that expected of a school prefect attempting to get some semblance of order from a class of new pupils. I had the misfortune in the later stages of my career to serve with a couple of officers like this. Having never been known to accept fools gladly, this is perhaps why, in June 1988, I took premature retirement after 34 years of Army service, 30 with the Intelligence Corps.

On retiring from Military Service, I became a civil servant, initially employed as a government investigation officer with the MoD (Air) then, on amalgamation with other vetting departments, what was to eventually become the Defence Vetting Agency and for the next fifteen years I was employed to look into the affairs and private lives of those people who were hoping for employment or already were employed in work of a sensitive nature that required their entire life, past and present, to be investigated prior to special clearances being granted, just as mine had been for the previous 30 years. After years of living the life of a mushroom – kept in the dark and fed on muck – I was set loose on the general public to discover how some of them lived. Although I found the work to be extremely interesting and rewarding, on some occasions, having spent four or five hours with an individual asking questions of a personal nature to include finances, religion, sexuality and politics, I would frequently question my own lifestyle – perhaps it was I who was not normal.

4

THE CORRIDORS LINKING EAST AND WEST

Before I launch into the main body of my story, my experiences during the Falklands War, I feel I should give a flavour of what it was like to serve in the British Army when at the mercy of superiors not quite up to the task. Obviously, most were excellent leaders, but we were sometimes let down, both before and during Operation *Corporate*. When I think of the outcome of some of the orders I received and other events during my Army career, where it was evident the originator lacked sufficient attention to detail, my mind wanders to an event that happened in the mid 1970s, when I was serving with an independent signal squadron in West Germany very near to the East/West German border. As the Tp Commander and Operations Officer of a troop with its specialist, highly classified equipment mounted on land-transportable platforms, then the only unit of its kind within the British Forces capable of deployment in a tactical role, I was tasked to deploy the troop to Berlin for a period of five weeks on specialist operational duties.

The deployment would prove no problem because for the previous couple of years the troop had spent most of its time on operational trials and tactical deployments and had become extremely proficient in its specialist role. However, I was a little apprehensive and concerned that travelling through East Germany (GDR) could be a cause for concern because this would be the first time the troop had ventured outside West Germany (FRG).

Since the end of the Second World War, travel along the corridor between West Germany through East Germany to the city of Berlin was never straightforward. From the inception of the Federal Republic of Germany

(Bundesrepublik Deutschland) founded in May 1949, and consisting of the American, British, Canadian and French Zones, but excluding American, British and French Zones of Berlin, West Berlin remained a free but occupied city, not entirely separate from the Federal Republic of Germany. After the construction of the Berlin Wall and an American/Soviet tank stand-off at Checkpoint Charlie in 1961, West Berlin became a de facto part of West Germany. West Berlin was geographically completely separated from the FRG and access was only permitted through strictly controlled checkpoints. It was not until 1971 that a Four Power agreement guaranteed access across East Germany to West Berlin.

The procedures for Allied Forces travelling between West Berlin and West Germany were perhaps less restrictive than those for civilian travellers. The prime reason for this was that the three Western Allies did not recognise the establishment in October 1949 of the self-declared socialist state, the German Democratic Republic. Therefore, any Allied servicemen travelling between West Germany and West Berlin, in the event of being stopped for questioning, apprehended or detained by the GDR Police or members of the Ministry of State Security – the Stasi – were instructed not to acknowledge their existence by responding to any questions, but to immediately request the presence of a Russian Army officer, who would eventually, it was hoped, bring the situation to a satisfactory conclusion. In fact, what would happen, say, in the event of a minor traffic accident, was that the GDR Police would acknowledge the request for the presence of a Russian Army officer, keep the detainee waiting for about an hour with the excuse that they were trying to contact a Russian officer, then allow the journey to be completed unhindered.

There are always exceptions to the rule, one being in 1979, when I was serving in West Berlin. A Corporal under my command was involved in a serious traffic accident when travelling by motorcycle from West Germany to West Berlin. If he had awaited the arrival of a Russian officer he would have been dead from a ruptured spleen and other damaged internal organs. Instead he accepted help from the GDR Police who rushed him to a hospital operating theatre in nearby Magdeburg Hospital, after which he eventually made a very good recovery. However, this was not without its problems. As an EW Operator it would have been necessary for him to have been subjected to various vetting procedures and security clearances to allow him regular access to highly classified information. The type of information and data he would have been privy to would have made him a prize catch for a potential enemy espionage organisation. Not for one minute would it have crossed my mind to believe that the Corporal was in any way disaffected and would consequently voluntarily impart anything of a classified nature. But, while lying in the hospital bed in Magdeburg, what sureties did the British military have that he was not given a substance to make him involuntarily part with secret information? Remember, this event took place during the Cold War when the games played were not strictly in accordance with the rules. Eventually the Corporal was given a Medical Discharge as a result of his injuries, so the problem of what the Army did with him on his return to normal duties did not arise.

In a similar vein, in the early 1970s, I was taken to a military hospital in England for the removal of my gallbladder; when I eventually came around from the general anaesthetic I was aware that there was a Warrant Officer from the Intelligence Corps in the recovery room, security cleared to the same levels as myself, who was with me from the end of the operation until I was returned to my ward; his sole purpose in being there was to ensure I did not unintentionally mention anything classified while not in full control of my faculties.

The quickest and 'less hassle' method of transportation for service personnel travelling between West Germany and West Berlin was by air. Arriving and departing from an Allied Forces-controlled airfield in Berlin avoided unnecessary questions from Soviet officials. However, with the increase in travellers considerable use was made of rail and road. All forces from the occupied zones of West Berlin had the use of their own dedicated rail service. The Americans arrived and departed from a rail station in the area of their controlled zone in West Germany, the French used a station in Paris.

The first use of rail travel by the British was in July 1945 when freight trains and separate passenger service (with Army escorts on board to prevent looting) were used by elements of the Control Commission Germany taking up their positions in the future British Sector of Berlin. The locomotive crew were initially professional railwaymen, serving members of the Royal Engineers. By 1947, 'the Berliner' had established a regular daily overnight passenger service (in both directions) between Berlin and Bad Oeynhausen/Herford, the then location of HQ BAOR where there were connections with the British Army troop trains from the Hook of Holland. Twice a week the train continued direct to the Hook. During the time of the Russian Blockade (24 June 1948 to 11 May 1949), all land and rail entry points into the Russian Zone of Occupation (East Germany) used by the

Allies temporarily ceased. Rail traffic resumed in May 1949 and in the early 1950s, the western terminus was cut back to Hanover.

In the early 1960s, the 'Berliner' service was reorganised and became a daytime train based on a daily shuttle out and back from Berlin to the town of Brunswick in West Germany, where the British had a transport movement control unit. From here Allied servicemen would report and board a train owned and commanded by the British Forces but drivers, firemen and engineers were sub-contracted from the West German Railway Authority. This train would depart Brunswick Station at approximately 4pm every day of the year including Christmas Day, for a return journey to West Berlin. The route taken would be via Helmstedt, a small town on the West and East German border. On leaving Helmstedt the 'Berliner' entered the 'corridor', an area of no-man's land containing wire fencing, minefields, lookout towers and guard dogs. Travelling only a very short distance into East Germany the train would stop at the first station at Marienborn. Here the engine pulling the train would be uncoupled, it and its West Germany driver, firemen and engineers would then return to West Germany while the train was coupled to an engine belonging to the East German Railway Authority who would also provide the driver, firemen and engineers. At this stage the carriages were locked and the train virtually sealed from the inside to prevent any attempt at breaking in. The train would have a nominated British officer as Train Commander, a Russian linguist to act as interpreter and a duty army guard force. On occasions, intelligence spotters were also carried; their duty was to visually observe and record along the route any Warsaw Pact country troop movement, the deployment of military installations and changes within strategic military compounds and barracks.

At the time the changeover of engines was taking place, the Train Commander accompanied by his Russian linguist would detrain and, along with several Russian Army officers, participate in a short introductory ceremony on the platform before retiring to an office for refreshments while the train's manifest of passengers and cargo was scrutinised and recorded. Rumour has it that above the table, set in the ceiling, where the checking of passengers' passports was carried out was a camera to record the personal details of those travelling. Once the Train Commander returned to the train, the train would once more proceed in the direction of West Berlin.

The British train was the most popular means of travel amongst the Allied Forces because it provided full restaurant facilities; the American and French trains had a limited snack bar and apart from a small bottle of wine with the evening meal on the French train, both were considered alcohol free. It must be remembered that during my experiences of travelling on the British troop train, relationships between NATO and Warsaw Pact countries were extremely strained; living conditions, extreme poverty and a shortage of food and other consumer goods made life in East Germany, compared to the West, very miserable indeed.

With that in mind, was it just by chance that a one-class single sitting of a 'Silver Service' three-course evening meal, served on white linen, with wine if required, by waiters dressed in evening dress style suits with bow ties and white gloves, was served just as the train pulled into Magdeburg Station? There was usually a delay at Magdeburg during which time several hundred East Germans would have walked up and down the platform, passing the restaurant car and observing how the British Army fed its troops. On a few occasions during this enforced stop, troops, particularly members of sports teams visiting Berlin to compete in competitions, had been known to illegally open the carriage doors to allow access to the platform where they requested the local East German young ladies to dance; or in the case of rugby teams in particular there was sometimes a 'mooning' session. Their fun was usually short-lived and they inevitably paid for their indiscretions when they retuned to their units.

This enforced stop at Magdeburg allowed the East German Police to rid the train of any unauthorised passengers and/or asylum seekers, in particular when the train was on its return to West Germany. The Soviet Union and East German authorities, in order to ensure that even though they had been given a guarantee that only bona fide passengers were travelling on the train, were of the opinion that seeing is believing. Therefore, the train was repositioned to a protected 'corridor' away from the main platforms to allow members of the security force to carry out their own external inspection. This 'corridor' had a high, thick wire-mesh fence along both sides of the track to prevent side access on or off in the siding, the only access was either forward or to the rear of the train. While in a stationary position in the 'corridor', high pressure boiling hot steam was released from pipes running parallel with the train track directed upwards in the direction of the bottom of the railway carriages. Anyone hitching a ride by suspending themselves from below the floor of the train carriages would, once the steam had been turned on, release their grip and beat a hasty retreat either forward or to the rear of the train. Once there, dogs with their handlers would be waiting for them. The train would stop once again at Potsdam to allow the East German engine to be detached and the East German rail employees to leave the train, before completing its journey and arriving at Charlottenburg Station, West Berlin, at 7.45pm, where it was cleaned and prepared for its return to Brunswick Station the following day.

After the train, the next most used route was by road. A German autobahn linking the port of Hamburg in West Germany to Berlin had been built before the Second World War and British service personnel travelling by car used this for access to West Berlin. Their journey started from Checkpoint Bravo located in the small town of Helmstedt on the border, where travellers booked in at the small Military Police detachment. Permission to travel was only authorised if in possession of a Berlin travel document and Forces identity card and/or passport. All travellers were timed on departure and on arrival. There were two reasons for this: if the driver exceeded the speed limit it would show through his early arrival at the Berlin Checkpoint, and if the driver had run into any unforeseen circumstances, his non arrival would initiate recovery procedures.

From the Military Police detachment the traveller proceeded through 'no-man's land', a piece of land across the border where movement other than by NATO travellers was not permitted and which was heavily guarded by armed East German and Soviet troops, both on the ground and in overhead towers. Having crossed, the traveller reported to the Soviet checkpoint for further documentation. Here, the driver and any passengers would have to leave their car for an inspection of their travel documents by a Russian Army officer; this could take from two minutes to two hours depending on the zealousness of the Russian duty personnel. It was only at the entry and departure to East German Soviet checkpoints (apart from the very few Allies involved with the incarceration of Rudolf Hess in Spandau Prison when the Russian Army provided the duty prison guard, and those Allied servicemen serving with the Berlin Air Joint Safety Unit) that British and Russian Armed Forces were able to engage in restricted but close verbal contact.

Although fraternisation between British and Russian troops was forbidden by Service Law, it did go on. This was evident by the slow introduction to the West of the *Shapka-ushanka*, 'ear flaps hat', a traditional Russian hat that since the winter of 1940 has formed part of the Russian soldiers' uniform. While awaiting the return of their travel documents the young British soldiers were often approached by young Russian soldiers who bartered various parts of their military uniform for pornographic magazines, then not available in the Soviet Union. Later the exchange was for cash with the going rate of a couple of West German Marks for a Soviet Army cap badge.

On completion of the documentation checks by Soviet officials at Helmstedt, the traveller continued his journey to the Russian exit checkpoint at Berlin, where once again documents were checked before proceeding through another highly guarded and protected 'no-man's land' before eventual entry into occupied West Berlin. Here the driver's duration of the journey was recorded and if arrival before the expected time was noted he paid a financial penalty for speeding through East Germany. The driver also had to complete a *proforma* relating to his en route experience and to report any contact with Soviet officials or servicemen.

As a representative of an intelligence collection agency holding the highest of personal security clearances and regarded as a good catch for any hostile organisation, my travel along the Berlin Corridor was initially more restricted than for most. When I first started to travel the corridor I could only do so by following the 'sweep'. Twice a day, morning and late afternoon, the Military Police would 'sweep' the corridor to ensure no Allied servicemen had been involved in traffic accidents, were experiencing mechanical problems or had been detained by the GDR Police. Later these restrictions were eased to allow me to travel in my vehicle but in tandem with another British serviceman. Eventually rules were relaxed to allow me freedom of travel along the corridor any time I wished.

Only once was I ever stopped by the GDR Police while travelling the corridor and that was on the grounds that my driver was exceeding the speed limit. Aware of my rights I requested the presence of a Russian officer. My driver at the time was an Intelligence Corps WO and fluent German speaker who overheard the conversation between the GDR Policeman and a colleague to the effect that although we were right to request the presence of a Russian officer, they would make no effort to request one on our behalf but instead would keep us waiting for an hour while they sat in their police car having a cigarette and cup of coffee. When the hour was up they informed us they could not contact a Russian officer and we were free to continue with our journey.

Returning to my experiences of the mid 1970s, serving with an independent signal squadron in West Germany, I was aware that my unit would have no input in deciding what form of transportation for our deployment to Berlin would be used. We were very pleased when notification was received that HQ BAOR had agreed, owing to the security implications, that the troop with its vehicle-borne equipment would be airlifted into the city. As time for the deployment approached, nothing had been heard regarding the flights and on making enquiries as to the delay I was informed that the early instructions for transportation had changed and the troop would have to drive its own vehicles through East Germany to Berlin.

As part of the deployment plan I was instructed to contact a signal squadron permanently located in West Berlin, to check if I would need any special external markings or insignia on the troop's vehicles to ensure they would give the impression that they blended in with the other vehicles already in use with British units serving in the city. In response I received a package containing self adhesive transfers of the Royal Signals' badge plus other informative markings, together with a picture of a Land Rover showing where the various decals/transfers should be placed. I discovered later that the Land Rover in the picture was that used by Her Majesty Queen Elizabeth II and other members of the Royal Family on their formal visits to inspect the troops stationed in West Berlin. It was a one-off and no other service vehicle in use in West Berlin was externally decorated like it. In the meantime, a staff officer, serving with HQ BAOR, decided that for security purposes, the troop's vehicle number plates should be changed so as not to identify them as belonging to a unit in West Germany.

Never having trained as a road transport movements officer, in the absence of official instructions I started to make my own enquiries as to how many vehicles constituted a convoy for passage through East Germany, just in case there were any restrictions on the size of convoys. I was amazed that such a simple question could be so difficult to answer. I was eventually informed by higher authority that a unit of ten vehicles or less did not constitute a convoy and therefore, did not require authorisation from the Russians prior to travel. As the British did not want to give the Russians prior warning of their intentions to send a small specialist troop to Berlin, it was decided that the troop would travel with a maximum number of ten vehicles as a group departing at the same time. Any additional vehicles would travel independently on the same date but at a later time.

On the day for the troop's deployment we left barracks under the cover of darkness as ordered. This precaution didn't make much sense because the troop regularly deployed in the direction of the border. The troop divided into a ten-vehicle convoy followed by four free runners (admin vehicles), and made its way to the rear of the NAAFI car park at Helmstedt, near the British/Russian checkpoint, for entry by road in and out of East Germany.

Here, still in darkness, the old vehicle number plates were removed and new ones, in pristine condition and in alphanumerical order, were fixed to the vehicles. The convoy advanced to the Helmstedt Royal Military Police checkpoint where, because they had been given prior notification of our planned deployment, formalities were quickly completed and in a short space of time we entered 'no man's land' and reached the Russian checkpoint for entry into East Germany (the other four vehicles were instructed to follow later). To my surprise the Soviet officers did not ask any awkward questions and with no apparent problems with our entry documentation the convoy set off through the 'corridor'.

At the end of the corridor all exit formalities with the Soviet authorities were completed without problems and the convoy entered West Berlin, completed Allied checkpoint formalities and made its way to the airfield at RAF Gatow where it was to establish its operational role.

Circumstances dictated that the troop had to establish its operational complex on the apron of the airfield runway, and though the complex was covered with camouflage netting, the local area was so open with no other forms of concealment or cover available, it was virtually impossible to hide the activities of the troop from prying eyes. The East/West Berlin border was only some 80 yards from our location to the border fence with no trees or other forms of natural concealment between the vehicles and the fence. To avoid our conversations being overheard, external speakers were erected through which music from the radio was played 24 hours a day. It was obvious from the glitter of binocular lenses on the other side of the border that the Russians observed the troop at all times.

After five weeks our operational duties were complete and the troop packed up to return to West Germany. On the morning of our departure from West Berlin, I was instructed to meet with the Garrison Duty Officer at the Berlin departure checkpoint who, together with a Russian interpreter, would conduct the troop's safe passage past the Berlin Russian checkpoint. What a mistake to think, like our outward journey, we were going to have an easy return journey. Leaving the Allied checkpoint my convoy of ten vehicles proceeded through the heavily guarded 'no man's land', along a small section of dual carriageway and came to a halt in a designated area outside the Soviet checkpoint, where I was joined by the Duty Officer and interpreter who had made their way independently. Leaving the parked vehicles the three of us made our way on foot towards a single-storey building where we were expecting to commence documentation procedures for entry into East Germany.

As we approached the building we observed three Russian Army officers standing at the entrance to the building and then suddenly the Garrison Duty Officer, who had been virtually silent until then, quietly muttered, 'Oh dear, Major Molotov is the duty officer, I thought he had completed his tour of duty and had returned to Russia, invariably when he is on duty he gives us trouble so I hope all your papers are correct.' Major Molotov and two Russian officers came towards us; I was introduced to him and from then on all proceedings were conducted in Russian through the British interpreter, who was a WO2 from the Parachute Regiment. All contact with the Russians was carried out in the open and I was never invited to enter the building where travellers normally conducted exit and entry formalities. The Soviet Major asked for my papers of authority to travel; these I handed to him, and after a limited scrutiny of the documents he glanced in the direction of my convoy and asked if the doors to the rear of the vehicles opened, to which I replied in the affirmative. When asked to open them I declined and at this stage the Russian began to get very irate.

After several refusals by me to open the rear door of the vehicles the Duty Officer, in an attempt to defuse the situation, took me to one side and informed me that under past agreements between the Soviets and the Allies, the Soviets had a right to inspect all vehicles proceeding along the East German autobahn. I informed the Duty Officer that even though he held a commission in the British Army, was the current Garrison Duty Officer and representative of the General Officer Commanding Berlin Forces, unless he met certain personal security criteria regarding access to very highly classified and sensitive data and other information, he too could not be privy to what was in the back of the troop's operational vehicles. During this 'standoff' the British interpreter was not exactly smiling because this was his last day of duty in Berlin; the following day he was to return to the UK on retirement from the Army. My refusal to comply with the Soviet officer's request raised the spectre of an extension of service to give evidence at my court martial. After a while the Soviet Major left us and entered a building; I observed him through the window making a number of telephone calls and in all he was in the building for about 20 minutes.

While he was away I tried to explain to the Duty Officer and the interpreter why I could not permit the doors to the vehicles to be opened. Deep in conversation I initially failed to notice that my convoy of ten vehicles had been completely surrounded by armed Russian soldiers; where they had come from I did not know but it was as if they appeared from a hole in the ground. Sensing that the situation could worsen the Duty Officer inquired what I was going to do now. I replied that I would have my ten vehicles cross over the central reservation, make a 'U' turn and return to West Berlin. He then asked what I intended doing if the Soviets blocked my return, to which I replied, 'I will worry about that when it happens.' Prior to the troop's deployment I was given a copy of Vehicle Movement Instructions as originated from HQ BAOR, but nowhere was it mentioned that the Soviets had a right to check the contents of Allied vehicles proceeding along the autobahn in and out of East Germany; had I been made aware of such a clause the troop would not have deployed by road in the first place. While contemplating my next actions the Soviet Officer returned, handed my papers back and informed me I could proceed on my journey. I didn't need to be told twice.

Having put several miles between the convoy and the checkpoint, I suddenly remembered that I might have to go through it all again at the Helmstedt checkpoint near the East/West German border – and how right I was. On arrival I brought the convoy to a halt in the inside lane of the dual carriageway to allow room for the four independent vehicles which, because of the unforeseen delay at the Berlin end of the corridor, could not have been far behind, to park alongside in the outside lane. At the checkpoint I was met by a Major Procter, Royal Military Police, whose attitude towards me, without knowing the full details of my altercation with the Soviet authorities at their Berlin checkpoint, was unwarranted to say the least. Major Procter produced a copy of a signal he had received from the Senior Signals Officer in Berlin, in which I was instructed to allow the Soviet authorities access to the back of my vehicles if requested, anything in the back of the vehicles of a sensitive or classified nature was to be covered with Major Procter providing the necessary covering as required.

I informed Major Procter that I could not carry out these orders and the Soviets would not have access to the rear of the vehicles. I was then asked what I would do if the Russians demanded access; I informed him that my Standard Operation Procedures required the troop to be self-contained for up to seven days. I would pitch a camp at the side of the road and would place an armed guard on the vehicles to prevent the Soviets' examination. At the mention of arms and ammunition Major Proctor nearly burst a blood vessel and wanted to know who allowed me to travel through East Germany with weapons and live ammunition. My response was that I was never informed of any instructions regarding the non carriage of weapons and/or ammunition and once again, because of the classification of equipments within the vehicles, if I had known about such instructions then the troop would not have deployed to Berlin without the means to safeguard the contents of the vehicles.

While engaged in conversation with Major Proctor several Soviet officers, one of whom was a Lieutenant Colonel, came out of a building to meet us; they were so close to us before we realised they were there that they must have heard part, if not all our conversation about travelling with weapons. This time there was no British–Russian interpreter; the Lieutenant Colonel introduced himself in perfect English. He first took my papers and handed them to a junior officer for inspection, then having established that I was the English Officer that had denied Major Molotov from the Soviet Berlin checkpoint access to the back of my vehicles, he invited me and Major Proctor to walk with him along the column of vehicles making up the convoy. While doing so the four free-running vehicles that left Berlin some considerable time after the troop's departure were observed approaching the checkpoint. As they got closer it was obvious from their movement from one lane to another that they didn't exactly know in which lane they should park. Eventually they decided not to park behind us, but instead parked alongside in a separate lane.

When the vehicles came to a stop the Lieutenant Colonel asked why the four vehicles had not parked behind my convoy; I explained they did not belong to my unit. He commented that this was strange as all fourteen vehicles had left West Germany on the same day and were returning on the same day; fourteen vehicles had been noted in the Helmstedt NAAFI car park changing their number plates, fourteen vehicles had been observed on the airfield at RAF Gatow and these fourteen vehicles were the only ones in Berlin decorated with identical decals, and now I was trying to explain that they did not belong to me. Then came the crunch question. Would I open the doors to the vehicles? I replied 'No'.

From my initial sighting of the Russian officers at Helmstedt all I could think was whether my refusal to open the vehicle doors was going to start an international incident similar to that of the American/Soviet Union tank standoff in 1961. On that occasion East German guards were authorised to examine the travel documents of a US diplomat passing through to East Berlin. After a standoff lasting five days, both the Americans and Russians called forward ten tanks each on either side of Checkpoint Charlie where they remained until the affair ended peacefully 24 hours later. It was obvious that Major Proctor had not been properly briefed regarding the movement and security implications governing specialist units such as mine, and was therefore limited in what he could say in my defence.

Having stood my ground, I was slowly resigning myself to the fact that the only solution to my predicament was to move all fourteen vehicles from the dual carriageway – which by this time had caused a build-up of other vehicles travelling in the same direction – establish a base under the protection of an armed guard, and pass the problem to those in HQ BAOR to resolve.

Having mulled it over in my mind and almost arrived at the decision that this was the right thing to do, I decided to change my approach; why I did it I will never know. I proceeded to explain to the Soviet officers that within the British Army Officer Corps there is an expression 'My word is my bond'. The Lieutenant Colonel replied that this was also the same in the Russian Army. I then informed him that I had nothing illegal in the contents of the back of the vehicles, I was not trying to hide stowaways and neither was I transporting illegal immigrants or political refugees out of the East. The troops I had with me were all visible through the windows of the vehicles and I was only travelling to and from Berlin on the instructions of a higher authority. Much to my surprise he thanked me for my cooperation, returned my papers and told me to proceed to my unit in West Germany.

Another example of those in higher authority failing to do what was necessary. I did not originate the content of the troop's Operation Instructions, nor make the decisions regarding who had access to the troop's operational data, the security regulations regarding equipment and its product and the use of arms to protect the product. These instructions and orders were issued and signed by others in the Army's command structure in compliance with policies originating at government level, particularly those relating to national security, which should have been known to those in the higher command structure responsible for issuing my orders and instructions.

At the time I was only a very small cog in a very large wheel but I have to ask why was it left to me to prevent what could have escalated into an international crisis? On the other hand, over the years I have wondered what if the outcome to my predicament had taken a different direction and resulted in just such a crisis. What stories I could have told when asked 'What did you do in the Army Grandad?'

5

THE SPECIAL TASK DETACHMENT

As mentioned in the previous chapter, at the end of 1981 I was posted from a small specialist signals squadron in Berlin to Coms & Sy Gp (UK), Garats Hay, Loughborough, Leicestershire, and assumed the appointment of Squadron 2i/c. Elements of similar organisations to this unit had been located at Garats Hay since the late 1940s, and an MI8 site manned predominately with civilian staff had been located close by at Beaumanor from the beginning of the Second World War. Until several years after the war, both the civilian and military sites coexisted as close neighbours.

At the time of this tour of duty at Garats Hay the barracks housed both Intelligence Corps and Royal Corps of Signals personnel undergoing specialist trade training for future employment in their respective corps as electronic warfare operators (EW Ops). Intelligence Corps personnel received training in foreign languages and analytical studies of active and passive features of foreign radio communication recognition procedures, while Royal Corps of Signals personnel, in addition to foreign language training, were also trained in the recognition and intercept techniques of foreign radio communication facilities. Coms and Sy Gp (UK), was also the home of a Special Projects Agency (SPA) where a number of highly trained and specialist qualified Royal Corps of Signals technicians were employed. Also accommodated in a secure area alongside SPA, was a small contingent of both Intelligence Corps and Royal Signals trained EW Ops, employed on the intercept and analysis of potential enemies' radio communication systems emanating from the communist countries in central and eastern

Europe. In addition to my specialist intelligence duties in command of this small strategic operational troop (Ops Tp), I also had command of a special task detachment (STD, a somewhat unfortunate acronym). The STD was administered by Coms & Sy Gp (UK), but the detachment, for operational purposes, was under direct control of DI24 (A), at the MoD.

The role of the STD was twofold; firstly, to trial new electronic and radio equipment for future issue and use within the Army's strategic Sigint and tactical EW organisations and secondly, as a small tactical EW quick reaction force to be deployed at short notice anywhere in the world. At this particular time, apart from the STD, the British Army's sole tactical EW resources were those belonging to 14 Signal Regiment (EW) stationed in BAOR, with the mission to provide Commander 1st British Corps with electronic support. Therefore, all non-BAOR areas of interest requiring tactical EW support would in theory, necessitate the deployment of the STD.

Manpower for the detachment's trials role was from the STD permanent cadre of myself, a WO1 Radio Supervisor and a Corporal EW Operator. However, when deployed in a tactical EW role or as a quick reaction force, the permanent cadre would have additional EW operators and linguist support determined by the size of the operation to be mounted and the geographical area of the target in question. As a World Wide Quick Reaction Force, the unit operated from within two air portable Land Rovers, linked by a purpose-made penthouse. Administrative and logistic support would be provided by the host organisation. The unit was expected to practise its air portability role regularly and to fulfil this requirement frequently attached itself to the Tactical Communications Wing (TCW), RAF Brize Norton, who on all such deployments readily accepted the role of host organisation.

On assuming command of the STD I was aware the Detachment's next Air Portability Training Exercise with TCW was scheduled for February 1982. With the Christmas and New Year break coming up it was not until January 1982 that I was able to get to grips with my new areas of responsibility. Therefore, not being fully conversant with my duties as Sqn 2i/c or those as the Ops Tp Commander, coupled with the fact that the Squadron Officer Commanding (OC) would be absent on leave during part of the scheduled training exercise, I decided that the main focus of the detachment's role would be on air portability training with very limited, if in fact any, EW role to be played. I would remain in Loughborough to familiarise myself with my new appointment. However, I was actively engaged in the planning and preparation of the detachment for its second deployment to Ascension Island (the first deployment had been in the autumn of 1981).

In its operational role as an Air Portable Unit the STD had no problems with the transportation of specialist operators, equipment or its own vehicles; however, owing to a lack of its own integral communication equipment, it did have a major problem with communications between its tactical location anywhere in the world and its strategic location (England). The principal reason for tactical deployments and co-location with the TCW was that in addition to its many other responsibilities, TCW had the responsibility for the Land Forces Long Haul Communication Systems, therefore being co-located made it easy for the STD to communicate electronically with others anywhere in the world.

During this February detachment to Ascension it was deemed necessary for crew members to practise their specialist skills and they were encouraged to use their, to some, unfamiliar equipment, in the search for possible target communications systems. Searching the radio spectrum for Morse and voice communication facilities of a non-commercial or civilian nature, they successfully intercepted several military and quasi military communication facilities transmitting from Africa. In addition, several ground-based military radio networks operating in the high frequency (HF) bands and believed to be located within Argentina were also intercepted.

On returning to its strategic location in Loughborough, the STD began to plan for its next overseas operational deployment, the location of which at this early stage was not confirmed, but an exercise deployment had been planned for a period between May and June of 1982. At this time neither we nor anyone else could have forecast that the next deployment would not be for training purposes but instead we would be departing on a 'war footing', ironically starting more or less from the last exercise location.

Much has been written and spoken about Operation *Corporate* (the name given to the British military operation to retake the Falkland Islands) during the past 30 years, and because of the sensitivity of some of the information on which my story is based, together with adherence to restraints imposed by the British Government's Public Records Acts, I have had to wait until now to reveal the crew of the STD's role in the Battle for the Falkland Islands.

With the passing of time, whilst the memory is able to retain masses of data, unfortunately my memory for people's names is not great. Because of this, and the fact that while I remember the names of the crew of the

STD and a few other people with whom I had very close contact, since my return from the Falkland Islands I have had no contact with the majority of people I served with during Operation *Corporate,* I have chosen, in most cases, not to reveal identities – they will all know to whom I refer by their deeds and actions!

Shortly after our return from the Falkland Islands and my eventual return to 'normal' duties at Coms & Sy Gp (UK), I was asked by my OC to give a presentation to the Senior Officers EW Course. The week-long course, held annually at Comms and Sy Gp (UK) was primarily designed as an insight into the world of EW for those senior Army officers about to be appointed to command Signal regiments or other specialist staff appointments. This I agreed to and when I asked how long the presentation should take, I was informed that I had a two-hour slot to fill! Having co-opted the Detachment WO1 to assist with the presentation, as the time got closer, although we had a good idea what we would say, I thought it best to have at least one rehearsal with an audience comprising the security-cleared training staff at Coms & Sy Gp (UK). As this was still in the days when Wednesday afternoons were deemed to be 'sports afternoons', it was quite simple to arrange a presentation for permanent staff while the trainees were participating in sport. At the end of the presentation, as usual, I asked if there were any questions and was met by a wall of silence. This was hardly encouraging.

When a member of my audience finally broke the ice and spoke, however, the reasons for the silence were revealed to be something other than boredom. They were shocked, not bored. I shall describe what the reasons were later. I will now tell the same story I told them in greater detail.

The early days. 2 Wireless Communication Squadron, otherwise know as Station X, based on Windy Ridge in Whaddon, Milton Keynes.

Thirteen apprentices after they were charged with 'mutiny', the author seated centre front, Harrogate, March 1956.

My visit to Belsen
concentration camp,
May 1958.

Three weeks spent
with the Canadian
Armed Forces in Arctic
conditions, 1971.

14 Signal Regiment (EW) Formation Day Parade, 1 July 1978. The Author is the second officer on the left.

The UK/US intercept station on Teufelsberg Hill in Berlin. It was used by a squadron parented by 13 Signal Regiment to report on the communications of Warsaw Pact countries.

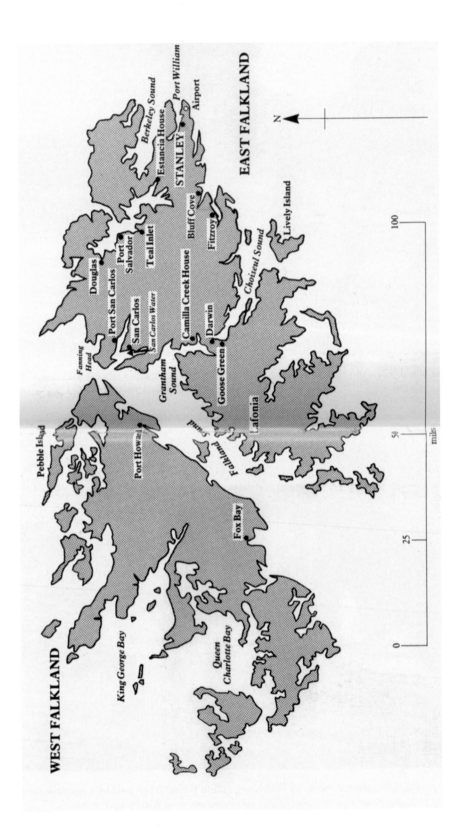

HMS *Intrepid* was the STD's Seaborne Operation Platform for 11 weeks during Operation *Corporate*. Launched 25 June 1964 and commissioned 11 March 1967, 12,120 tons, length 158.8m, beam 24.4m, speed 21 knots.

HMS *Fearless*, a LPD and sister ship to HMS *Intrepid*. It housed Marine Y Troop and was the location of 3 Brigade HQ during the Falklands War.

HMS *Hermes*, the flagship of Rear Admiral J.F. Woodward during the Falklands War.

Gurkhas of 5 Brigade keeping fit on board the *QE2*. On arriving at Port Stanley they, unlike the Scots and Welsh Guards, were fit enough to 'yomp' to relieve 2 Para, while the Guards waited for transportation.

SS *Canberra*, which transported the Scots and Welsh Guards from Grytviken, South Georgia to the Falkland Islands.

The *QE2* which transported 5 Brigade from the UK to Grytviken, South Georgia, arriving on 27 May 1982.

Arrowflex machine used to encrypt reports before transmission.

Sights from HMS *Intrepid* during
Operation *Corporate*.

HMS *Antelope* sinks on 24 May,
the day after being attacked by
four A4-B Skyhawks of Grupo 5.

An unknown vessel and a view of Port Stanley as seen from a landing craft.

HMS *Conqueror*, the submarine which sunk the ARA *General Belgrano* on 2 May 1982, passing HMS *Penelope*. The author was later tasked with producing a report on the sinking, in which 323 lives were lost.

The *Sir Galahad* burns the day after being bombed in Bluff Cove on 8 June 1982.

A Harrier GR3 lands on HMS *Hermes*.

Fort Grange replenishes *Iris* by jackstay transfer.

Argentinian PoWs wait for transfer from Port Howard, West Island to HMS *Intrepid*, 15 June 1982.

The PoWs are embarked on HMS *Intrepid*. Many were conscripts and had suffered food shortages during the campaign.

Port Howard on 15 June 1982.

The STD arrive in Port Howard to find the Union Flag flying once again, 15 June 1982.

A Royal Engineers Sapper lifts an Argentinian anti-tank mine. The Islands were extensively booby-trapped, making travel very hazardous even after the cessation of hostilities.

The Prince of Wales aboard SS *Canberra* welcomes returning officers, 12 July 1982.

From: Colonel C J Gilbert FBIM M Inst AM

MINISTRY OF DEFENCE Room 0156

Main Building Whitehall London SW1A 2HB

Telephone 01-218 3587 (Direct Dialling)

01-218 9000 (Switchboard)

Captain D J Thorp Intelligence Corps	Your reference
JSSU FI	Our reference
c/o HQ LFFI	25/3/4 Encl 1
British Forces Post Office 666	Date
	8 July 1982

Dear David,

I am writing this letter to you now in the hope that it reaches you in reasonable time before your departure.

Whilst the fighting as such is over, the efforts of the organisation are really only just beginning with the establishment of the permanent station. The burden of this is inevitably falling on your shoulders and those of Mike Dawkins and I am sure you appreciate that to ensure that we set up an efficient operational station we have had to rely on the expertise of our "resident" Siginters.

Looking back over the past 3 months or so, whilst I personally have not been privy to your product, I have heard nothing but praise from the customers for the prodigious efforts put in by you and your detachments. You had little time to get organised and mounted, were thrown into a strange environment and had little guidance on your modus operandi. That the detachments were so successful is due in no small measure to your professional approach, enthusiasm and, above all, your initiative. I fully appreciate that all this was carried out in a very dangerous and hazardous environment, in adverse living conditions and foul weather.

You and your detachments have made a significant contribution to the overall intelligence effort and by it enhanced the reputation of the organisation.

I most sincerely congratulate you and your team on a very difficult job carried out with great skill and application.

Finally be assured that I am doing all in my power to have your detachment replaced as soon as possible so that you can be re-united with your families.

Yours aye

Colin

The Author (second from right) receives the 'Intelligence Prize', 1983.

David Thorp today.

6

PREPARING FOR WAR

It was Tuesday 13 April 1982, the day following the Easter Monday, and my two younger children were home from boarding school and I had taken my youngest son for a round of golf. On our return home I was informed by my daughter that the Adjutant from Coms & Sy Gp (UK) had left a message for me to call him. I returned the phone call only to find out that the remainder of my leave had been cancelled and I was to return to my unit reporting for duty first thing the following day. On arrival at Garats Hay, I was informed by the Adjutant Captain Bill Clark, Royal Signals that the STD, with me in command, was on immediate 'stand by' for deployment to the South Atlantic. Exactly where within the area of the South Atlantic and when was at that time not known. I was also informed that while I had been enjoying my Easter leave, other members of Coms & Sy Gp (UK) and DI24 (A) had been busy locating and arranging the availability of additional manpower with the necessary specialist skills to support the STD against a Spanish-speaking target operating within the area of the South Atlantic. In my absence it had been arranged for me to meet some new members of my detachment in the afternoon and also for me to attend meetings at HMS *Warrior*, Northwood, on 22 April 1982, and on Friday 23 April I was to attend specialist meetings at the Government Communication Headquarters (GCHQ), Cheltenham.

It was confirmed that the STD 2i/c would be the Supervisor Radio held on the detachment's permanent strength, in this case WO1, Royal Signals Allan Newman – a more suitable or better qualified person for the job would, I believe, have been very hard to find. We had both served together in Cyprus in the early 1970s and during the intervening years had had irregular contact. The epitome of a professional WO1, his personal qualities also made him a very likeable person, with the ability and expertise to turn his hand to anything with very creditable results.

Some of the crew members, because of the distance they had to travel to Garats Hay at such short notice, would not be present to attend the pre-arranged initial meeting of the specialist operators selected for this, our first non training exercise. However, the initial meeting went ahead as planned by WO1 Al Newman; those not present could catch up on the detail later. On entering the secure compound that housed the STD operational buildings and vehicles and where I had arranged to meet some of the members of the detachment, I observed a very large person with shoulder-length hair and a 'Pancho Gonzales' moustache. I enquired of a member of staff what the 'Mexican Bandito' was doing on the inside of a secure area and was informed that this was WO2, Intelligence Corps, Andy Gray, who had only that week completed a Spanish language degree course at York University. A very experienced field operator who had spent considerable time observing targets within Latin America, his language aptitude, both written and spoken, was excellent. Once he had 'scrubbed-up' and shaken off the professional student image, as large as he was, he turned out to be worth his weight in gold and helped greatly in the eventual overall success of the detachment.

The remaining four members of the detachment were Corporal Reehal and Corporal Lukehurst, both special operators, Royal Signals, and Corporal Gary Littlewood and Corporal Jock Cairns, both Spanish linguists, Intelligence Corps. All four were extremely competent operators with considerable experience within their specialist areas. In retrospect I was given a team of six individuals all at the pinnacle of their professions who, with very little effort, were able to mould themselves into a team that achieved results of the highest quality and proved themselves to be excellent ambassadors for the British Army and their parent organisations.

During the next four or five days, all crew members had reported for duty at Garats Hay, and the detachment's time was spent on familiarisation training on the specialist radio communication equipments and facilities available to the Argentine Armed Forces, as well as unique communication procedures available to them. In addition, when time permitted, they were coming to grips with the intricacies of their own equipments housed in the detachment's two Land Rovers.

On arrival at HMS *Warrior* on 22 April, I was introduced to the movements cell responsible for, amongst other things, finding space on ships heading south to transport vehicles. This movements cell was staffed by serving member of the Royal Marines. This was to be my first experience of working alongside the Royal Marines and, as I was to discover, working closely with the Marines is no different to working with any other branch of the British Armed Forces. As a fighting force they have my utmost respect and impressed me with their dedication to duty, enthusiasm and professionalism. In typical fashion and in keeping with tradition, the movements cell had not been informed through their chain of command of the inclusion of the STD within the Task Force, or to be prepared to accept two additional Land Rovers for the journey south. I was informed that I could not take my operational vehicles with me; they would however, be prepared to accept an additional seven bodies on what at that time was an unidentified naval ship. My immediate response was to telephone my parent organisation and MoD sponsor branch to inform them that, due to lack of vehicle space on ships heading south, we would not be part of the Task Force.

As previously mentioned the STD was under the direct command and control of DI24 (A), a department within the MoD, so it came as no surprise that shortly after my initial telephone call, the first instruction not allowing the transportation of my operational vehicles was cancelled; space had miraculously been found for not only the detachment crew but also the vehicles for the outward journey, at least as far as Ascension Island.

Having confirmed that we would now definitely be included as part of the Task Force, it was tentatively agreed between myself and the Task Force operational planners that the role of the STD would initially be to provide EW in a passive role. This meant no jamming, intrusion or direction finding, in support of operations carried out by 3 Brigade Land Force Commander. Furthermore, the STD were to carry out this role under the command of the Royal Marines Y Troop (Y Tp).

The Y prefix in a military organisation was originally used during the First World War and for a few years after in the context of organisations directly connected to Military Intelligence Department 8 (MI8). Throughout the Second World War and for a limited period after, the War Office Directorate of Military Intelligence comprised the following departments:

Department number and area of responsibility

1 Code breaking
2 Russian and Scandinavian Intelligence
3 Eastern Europe
4 Aerial Reconnaissance
5 Military Intelligence
6 Secret Intelligence Service
7 Propaganda and Censorship
8 Sigint Intelligence or Y Service
9 Resistance Workers and PoWs
10 Weapons Technical and Analysis
11 Field Intelligence Police
12 Censorship
13 Not allocated
14 Germany
15 Aerial Photography
16 Scientific Intelligence
17 Secretarial to other departments
18 Not allocated
19 Enemy Prisoners of War

Most of these departments became defunct immediately at the end of the Second World War, the exceptions being MI5 and MI6, which continue to be used colloquially for their respective agencies. Like MI5 and MI6, the term MI8 continued in use until the War Office became the Ministry of Defence and DI24 became the department with responsibilities for Sigint, Elint and Comsec. DI24 was further subdivided into departments with special responsibilities for Navy, Army and Air Force – the Royal Marines, other than through the Royal Navy, were never included as part of the old MI8 or its successor DI24; perhaps this is why they hung on to the old Y prefix.

By the time I assumed command of the STD, this very small Marine unit had been operational for several years, during which time the unit had renewed and updated both vehicles and equipment from those previously belonging to the STD. Though superseded by advanced technology they were perfectly serviceable and with all components repaired and refurbished before being passed to the Y Tp.

Manpower for the Y Troop was drawn from within the Marines. A few members of Y Tp had received specialist training at DI24-controlled units and were aware of activities in the wider circle of Sigint and Comint, and the majority were familiar with Comsec. Comsec as defined by the activities of Y Tp prior to April 1982, was mainly restricted to monitoring Marine exercise communications and reporting any obvious or apparent breaches of security.

On the conclusion of my visit to Northwood I was not overly impressed with what the Task Force HQ had planned for the role of the STD. The STD, an independent unit, was to be under the command of an organisation with out-of-date equipment, specialist operators not fully trained in the role of Sigint and limited language ability for the target area. However, possibly due to the timely intervention by DI24 early in my arrival at the movements cell, I was informed prior to my departure that the STD was to report to Portland dockyard for embarkation on the Landing Platform and Dock (LPD) HMS *Intrepid* on Saturday 24 April 1982, only two days after this briefing, with a ready-to-sail date of 26 April.

With all administrative and operational preparations completed by Thursday 22 April, the unit was ready for a lengthy deployment and I was able to leave Garats Hay for my family home in Cheltenham to ensure I arrived in good time for my briefings at GCHQ on the Friday. Being no stranger to GCHQ, prior to my arrival I was able to ensure that I was briefed by all departments with a current operational interest in the target area. This typically meant visiting a number of different departments and offices within each single department. Each department I visited went to great lengths to ensure they supplied me with a mountain of highly classified material, all in paper form, in the belief that it would be vitally important to the success of the operational role of the STD.

At about 4pm and feeling a little 'punch-drunk' from the saturation of all the information passed to me, I decided the time had come for me to take my leave. However, as was normal for about this time on a Friday afternoon, there were very few civil servants left at their desks as most were in the habit of accruing hours off through flexi working and took time off at the end of the week. My arms were full of classified material, but there was no one present able to advise me as to how I was to transfer all this information to Portland dockyard the following day. Leaving it to the following day would not have helped solve my problem because being a Saturday, even fewer employees would be working. So in the face of adversity there was only one thing to do: search around for a few plastic carrier bags, fill them with classified material, none of which I had signed for, depart through the main gate past security with fingers crossed that they wouldn't suddenly decide to do a bag search, and walk back to my house, which was only a quarter of a mile away.

On reaching my home it then struck me that I now had a serious security problem on my hands: all these classified documents but no recognised secure storage. Throughout the evening I never let the plastic bags out of my sight and when I went to bed I stored them under my bed. I hardly slept a wink for fear that of all nights, that was the night somebody would choose to burgle us.

Transport to take me to Portland dockyard arrived at my home early on the Saturday; I was relieved to see a driver in Army uniform because I now had an escort who would stay with me until I, with my two plastic bags, was secure in a military establishment. On saying my farewells to my wife and family I told them not to worry because it was highly likely that we would arrive at the Falkland Islands, circumnavigate them a couple of times and then return to England. How wrong I was!

Those plastic bags highlight the fact that sometimes risks have got to be taken; failure to do so could have jeopardised the operational role of the STD as it would not have had current intelligence data relating to its anticipated target. Whilst some were only too keen for me to accept all they had to offer in the way of, to them, extremely useful information, I only accepted that which through years of experience I knew to have a bearing and useful purpose, thereby reducing my holdings of classified information to the absolute minimum.

Arriving at HMS *Intrepid's* berth at Portland dockyard, I struggled up the gangplank with all my kit and the two bags. When I got to reception I was advised by the duty staff to report to the Commander who at that time was in the ward room. It was suggested that I leave all my baggage at the reception station until such time that I knew which bunk I had been allocated. A sensible suggestion and I left my kit but retained the bags – no amount of persuading by the duty staff would make me leave them. I have since often wondered what the Navy thought of an Army Officer going to war with his belongings in two plastic bags!

During my briefings at HMS *Warrior* three days earlier, I was informed that the one person on HMS *Intrepid* who would most certainly have the correct level of personal security clearances to allow me to discuss my specialist role or any problems I might have, would be the Captain. On my arrival, the Captain was absent on briefings and was not expected to return to his ship until late Sunday afternoon. By this time I was rapidly coming to the conclusion that I was never going to be free of the bags filled with highly classified material. However, on discussing the problem of storing it with the Commander (Ships 2i/c), he suggested I had a word with the ship's Communications Officer as it was extremely likely that he would have the necessary security clearances to help me. Fortunately he had the required clearances to handle the documents I was carrying and he was able to find temporary suitable storage until such time that I could make more permanent arrangements.

On his return from his briefings I was introduced to the Captain, who, being aware of the STD's need to go operational at the earliest opportunity, allowed us to set up an operational facility in the ship's EW room. This proved to be extremely useful because we had insufficient space to erect suitable aerials for coverage of the target area and, as we were to discover later, it would have been impossible to operate from within our vehicles on the ship's open decks as these were to be used for landing platforms for fixed wing and rotary aircraft. In addition to having warm, dry and fairly comfortable accommodation in the EW room, we also had access to an aerial distribution panel, which allowed the radio equipment to be connected to the vast array of ship's aerials. On Monday 27 April 1982, the assault ship HMS *Intrepid* set sail from Portland dockyard on her journey south with the STD fully established in operational mode.

The first couple of weeks on our journey south were operationally not very productive, although the intercept radio facilities were manned and operational 24 hours a day. I was aware the distance between the designated target area and ourselves was too great even for transmissions in the HF bands to be received, but what encouraged me to adopt this permanent operational posture immediately HMS *Intrepid* set sail was the fact that the unexpected could happen and from experience I was aware that through atmospheric and spurious transmissions, radio signals have been known to travel over vast distances, especially over water. I remembered that as an apprentice in Harrogate, during a local exercise operating the main battlefield radio, the No 19 Set with a range of approximately 20–25 miles we were able through spurious transmissions and a very good quality signal to communicate with a British Army signals unit stationed in Hong Kong, some 6,000 miles away.

The detachment remained in good spirits and morale was high. The lack of target acquisition allowed the crew of the STD to study, from the limited data available, our expected target array. Some familiarised themselves with their new working environment while others utilised the time to ensure they completed additional outstanding admin arrangements.

The issue of Arctic-type clothing to troops deployed to extremely cold climates is normally done under prior notice from a central depot before departure. In the short time from being placed on an operational footing to actually departing, the STD crew had insufficient time to be properly equipped with the full range of winter warfare clothing. This was where WO1 Newman came into his own. It was not long before he had made the necessary contacts with the Marines quartermaster department, had us all measured up and placed our order for the clothing and equipment to be ready for collection on arrival at Ascension Island.

Ascension lies in the southern hemisphere roughly equidistant from Africa and South America and about 700 miles north of St Helena. First discovered in 1501, it remained uninhabited until Napoleon was incarcerated on its neighbour in 1815. Ascension together with the islands of St Helena and Tristan da Cunha are grouped as a British overseas territory. The island does not have an indigenous population, with the majority of the inhabitants migrating primarily from St Helena, known locally as Saints, with a smaller transient population from Great Britain and the US. Its climate is drier than expected given its tropical location; the temperature is generally in the eighties Fahrenheit and rain is virtually unknown. However, over the years the island has become self sufficient in water and electricity to power refrigeration and air-conditioning units, so that even the hottest periods become bearable. Since a submarine cable was landed on the island in 1899, it has been a crossroads for international communications, a role that continues today with satellite and other forms of sophisticated communication equipment. For many years the island has been shared with the US who found it strategically placed as a tracking site for their space exploration program. The island has always been utilised as a supply base for naval ships but never to the extent or capacity that it was used during Operation *Corporate*.

As previously mentioned, the STD was established as an air portable unit with external communications systems between the detachment and its parent organisation and the deployed headquarters under whose command we happened to be, provided by a host organisation. In the past it was the RAF, now it was the Royal Navy. It was required that one of our Royal Signals special operators be trained in offline cryptography techniques to ensure that STD's product would not be compromised during transmission and, by completing its own basic preparation of message processing and despatch, this would avoid the problem being passed to someone else.

Prior to the February exercise deployment with TCW to Ascension, the STD had taken delivery for the purpose of evaluating field trials and feasibility studies, of three of the latest state-of-the-art offline encryption machines known as 'Arrowflex' offline encrypt/decrypt units. The Arrowflex resembled a large teleprinter with a 5-unit Murray code perforated tape reader attached. The unit was used as a normal typewriter, producing both a page print and perforated tape copy. Where it differed was in the built-in encrypt/decrypt device; by using a prearranged daily changing key setting, when data was typed into the Arrowflex it produced a readable page copy identical to the perforated tape but typed in print to make the reading of the tape more user-friendly. However, the perforated tape was encrypted and appeared as randomly selected figures and letters, displayed in blocks of five figures and/or letters. To decipher, the encrypted perforated tape was put through an Arrowflex with the same key settings and a plain language copy was produced in paper format through its integral teleprinter unit.

The STD deployed with two of the Arrowflex machines while the third machine remained at Garats Hay; this way, so long as we had a secure communication link we could communicate with our parent unit and sponsors via an extremely secure means – although the perforated tape may have contained highly sensitive codeword information it was treated as non classified until a page print had been produced and therefore, could be handled by non-cleared security personnel until reaching its final destination. Also, the physical handling of this semi-processed product was such that it could be transported by hand from the STD operational cell to the host unit providing the communications facilities without contravening security regulations, when and if the communications facilitates were detached and not co-located with the STD cell.

As standard operating procedures, the STD always carried a small quantity of perforated tape with the bulk of its requirement supplied by its host organisation, in the past the TCW. During the first week or so at sea I realised that the Operation *Corporate* deployment could be longer than anticipated and we would need more perforated paper than we had expected; a stock check was carried out. It revealed that we had insufficient stocks of tape for a prolonged period of time. Unless an additional supply was available in the foreseeable future the detachment would not be able to fulfil its operational role. A request was made to the communications section on HMS *Intrepid* to provide any future stock, but it quickly became clear that we had a major problem. *Intrepid's* communication section had deployed with a very limited stock of perforated tape, sufficient only for their own use. In fact, aware that the STD frequently asked *Intrepid* to transmit messages that included prepared perforated tape, the Royal Navy were about to ask us for some of our stock. It was apparent that there was an acute shortage of this commodity amongst all units heading south, and in a war zone the STD would have found it almost impossible to communicate with the UK without perforated tape or other forms of secure communications.

On 12 May, the STD intercepted its first Argentinian military communications network. Activity was in the HF band, in plain language (not encoded/encrypted) and of low grade intelligence; however, it did confirm that we were in a position to play a fully operational role when and if the occasion arose. Prior to our arrival at Ascension, one of our UK sponsors informed us of their interest in a radio communications network emanating from the Argentine military headquarters in Buenos Aires to the 'Malvinas' Force Headquarters in Stanley, Falkland Islands, and requested technical assistance from the STD in the capture and recording of this radio network. The major problem with this request was that although we could intercept the signal, we did not have the facility to capture its content. However, I was informed by our parent unit that equipment required to capture the transmissions would be despatched from the UK by air to a particular government facility on Ascension.

On my arrival I was requested to meet with an officer from this facility and take delivery of the equipment. However, on meeting the officer he informed me that the equipment had not arrived even though it was past its delivery date. Accepting there was nothing that could be done, I cheekily asked him if he had any spare perforated tape as the Royal Navy had set sail without an adequate supply! He said it was not a thing his station had much use for but perhaps he knew a man that could help. This man turned out to be the head of the US tracking station at Wideawake, and arrangements were made for me to meet with him. When I explained my predicament, he took me to a large prefabricated building which contained hundreds of boxes of perforated tape; the quantity I was to be given would depend on the date of his next delivery from the US. A quick phone call to his headquarters established that he could be re-supplied within a couple of days and told me to take as much as I wanted. I hurriedly arranged for the use of a vehicle with the help of the officer from the British facility and had hundreds of boxes delivered to the dockside for loading and distribution to the British fleet.

During the period of cross decking while at anchorage, I attended an operational meeting with the Marine Y Tp on HMS *Fearless*, also a LPD and sister ship to HMS *Intrepid*, to discuss 'plans of action and responsibility' in the event of a landing on the Falkland Islands. On board *Fearless* I was introduced to the 3 Brigade, Marine Corps G3 Int Captain Rose, Royal Marines, Captain Jerry Corbett Royal Corps of Signals (we had previously served together at 14 Signal Regiment (EW)) and Marine Warrant Officer Oxendine.

From the start of our prearranged meeting on board HMS *Fearless* it was obvious the Marines Y Tp had decided a plan of action without bothering to consult me or any other competent Sigint organisation and this, to say the least, made me a little irate. Why bother to extend an invite when everything was virtually *fait accompli* before my arrival? Since leaving Portsmouth, the Y Tp had not established the basic operational mode in which to familiarise themselves with the possible target communications systems they were most likely to encounter, in the belief that it would be alright to leave this until landing on the Falkland Islands.

This lack of forward planning was not entirely their own fault. Unlike the STD they were not offered the use of a ship's EW cabin and in common with the STD, were unable to park their vehicles on the open decks to enable them to establish an operational working area. It was obvious from this meeting that their Sigint expertise and linguistic capability was extremely limited; none of the Tp was capable of transcribing Morse code at a reasonable speed, their Spanish linguistic support was minimal and with some 25 years of continuous experience in Sigint operations and senior in rank to Captain Jerry Corbett, I was to be his deputy. I was also made aware that the Marines had allowed a civilian electronics engineer to travel under command of the Y Tp to allow him to conduct field trials on a piece of communications jamming equipment that he had developed – mind boggling to think that someone had considered it feasible to conduct field trials on new equipment during a war. There was to be more of this once hostilities started.

Not in a happy frame of mind I returned to HMS *Intrepid* to inform my detachment of the outcome of the meeting. The one redeeming factor was that I had persuaded Y Tp that because the STD was up and running with a well established operational, though small, organisation within *Intrepid's* EW cell, we should remain on *Intrepid* until the Y Tp had disembarked their personnel and vehicles, and were in a fully operational position to take over from us without any loss of continuity. Once this was established we would temporarily cease operations, disembark *Intrepid* and join up with the Y Tp on land. With hindsight this proved to be my best suggestion during Operation *Corporate*.

The few days spent at anchorage close to Ascension allowed the STD crew members time to relax and attend to some of the personal admin chores they had been unable to complete since leaving the UK. After the cold and damp of the English weather, the sunshine and warmth was such that it was a pleasure to be outside on deck rather than cooped up in a large metal trunk with only artificial lighting. The crew were encouraged to take advantage of their surroundings before moving south, where the weather was expected to become very harsh.

Several of the crew managed to get onto the island and for those who had been there only two months previously, it was a chance to reacquaint themselves with the place. The atmosphere on board HMS *Intrepid* became relaxed while personnel maintained an air of extreme professionalism. Surprising what a few hours of sunshine can do to lift one's spirits.

A couple of days after our departure from Ascension there was a knock on the EW cell door and there was a naval rating carrying a large package. He asked if I was part of a particular signals unit, and when I replied in the negative he asked if I had any suggestions where he might find the addressee. On taking a closer look at the package, although I didn't have any knowledge of the first part of the address, I was able to make out that whatever this unit may have been called, it was located at Garats Hay, Leicestershire, therefore in all probability it had reached its destination. I asked the rating how this package had come to be in his possession; he replied that someone had seen the package lying on the jetty at Ascension. It had been there for a couple of days, so when eventually the last Task Force landing craft (which happened to belong to HMS *Intrepid*) was about to make its final return to ship, this rating suggested the package be taken on board for onward delivery. On opening the package I discovered this was the piece of equipment I had hoped to take delivery of from the UK via the British facility on Ascension Island – the piece of equipment in question had a security classification of 'Secret', while the product it produced was classified 'Top Secret Codeword'. Regulations on the carriage and handling of such a package were that at all times the package should have been kept in a secure container and when transferred from one temporary custodian to another, a record including the temporary recipient's name, address and signature should always be obtained. This package had left the UK and sat on a jetty at Ascension, where it remained unattended for a couple of days, with absolutely no record of its journey. Even when it reached its final destination there was no official paperwork to signify that it had been received by its rightful addressee. Not exactly standard procedure.

7

THE TRANSITION FROM PEACE TO WAR

By the time HMS *Intrepid* had left Ascension Island, her embarked force that comprised mainly units of the Paratroop regiments and Royal Marine Commando had established a well-defined specialist to arm training program and was put through a rigorous daily training of live shooting (targets on tow at the rear of *Intrepid*), weapons drill, physical fitness, enemy recognition and identification and many other subjects. What struck me immediately were the dedication and enthusiasm members of the force exhibited in carrying out this daily training. Morale was extremely high, the troops maintained a good sense of humour and all displayed a very professional approach to the job in hand.

Within the STD events were progressing well; on Ascension the unit had received the full issue of Arctic warfare clothing and equipment. Operationally we worked a two-shift cycle of three operators working eight hours on and eight hours off with myself on extended days. We had with great difficulty installed the classified piece of equipment belatedly received on Ascension, and from 12 May, the number of different target transmissions available to our operation was increasing daily.

As a break from operations, I had arranged with a Navy Lieutenant Engineer to be given a tour of HMS *Intrepid*. This tour lasted all day, at the end of which I doubt there was anyhere on the ship I had not visited and I was informed by my guide that I had seen more of it than a great many ship's officers. After this tour, when I heard the call to action stations, and now being aware of some of the locations the sailors had to report to for duty in the event of an air raid, I was full of admiration for their courage. One such duty station was at the very bottom of the ship, accessed by a vertical ladder, where a solitary sailor would manually control the ship's rudder in the event of failure of the remote automated operation system located on the bridge. This duty sailor was totally isolated from the rest of the ship with only a telephone connected to the bridge for use in an emergency. The area was in total darkness, and the man would only be able to hear guns firing and bombs exploding, letting his imagination do the rest.

The STD personnel had by this time integrated well into the ship's routine, and had mastered some of the Navy 'slang', such as 'pusser's dust' (coffee), 'babies' heads' (individual steak and kidney puddings), 'one all round' (smoke break – originally from the days of Admiral Nelson when sailors on action stations were allowed a break for a smoke during a lull in the fighting, one smoke shared amongst them) and 'heads' (lavatories). We also had our vocabulary increased by one word contributed by the Marine Commandos – 'biff'; by the conclusion of hostilities many a Marine was shortlisted to 'biff' for England, having spent considerable time practising in the area set aside for prayer meetings.

The STD shift changeover timings were designed to fit those of the ship's watch changes. As it was normal practice for the duty watch to walk in front of the Embarked Force for quick service at meal times, it was not long before detachment members were invited to do likewise. This small gesture of cooperation proved to be the first instance where personnel of the STD began to be treated as ship's crew rather than a foreign force.

Sailing in the South Atlantic during winter is not something I would recommend. HMS *Intrepid* was not the smallest of ships, she was of a weight and size that enabled her to maintain buoyancy in most rough or choppy waters, giving the appearance of floating on top of the waves. However, when in adverse weather conditions with waves in excess of ten metres or so, she would be virtually submerged. As most waves at the time were over ten metres, we seemed to be perpetually underwater. Coupled with extreme cold and gale-force winds, the voyage was not a pleasant experience. Fortunately, none of the crew of the STD was adversely affected by the extreme weather conditions and absenteeism through sea sickness was never a problem.

The first couple of weeks sailing on a warship was, for me, relatively comfortable; not quite up to the cruising standards of a Cunard liner but I was, for most of the time, the only occupant in a four-berth cabin complete with bunk beds and proper mattresses (not hammocks), a sink with hot and cold running water and bath and shower facilities close by. The ward room stretched across the width of the beam, divided into two by curtains arranged to form a corridor, forward and aft, to separate the dining area from the bar-lounge area. Simply decorated with pictures and pieces of naval memorabilia, it was a pleasant place to escape for a little peace and quiet at meal times. High standards of mess etiquette were maintained with a choice of more than one dish at each meal time. While the service given by the mess staff, all of whom had secondary duties, was not silver service, all meals arrived ready plated, were hot and appetising. However, this was to change.

On leaving Ascension Island, orders were received for all ships to prepare for war; amongst other things this meant 'decluttering' the ward rooms. The corridor of curtains, furniture and wall coverings manufactured from inflammable materials were removed, mostly into store but some items were thrown overboard, leaving a large expanse of space with a few dining tables and chairs. After this makeover, apart from officers taking a hurried meal at the few remaining chairs and tables, the ward room was hardly used until the cessation of hostilities. The reason for taking quick breaks was that ship's manning levels were so fine-tuned personnel could only spend the minimum amount of time away from their posts. Also, the ward room stewards had been taken from their action stations to provide 'waiter' service in order to speed up the time taken for meal breaks.

As part of the decluttering exercise, all ships officers, including myself, became custodians of the bar stock and each was given a barrel or cask of beer to look after – this was to enable the bar stock room to be used for other purposes. By this time it was also found necessary to conserve water; consequently all baths, showers and a limited number of wash hand basins had their taps removed to ensure they were unserviceable.

Between the period of exercising with TCW on Ascension Island and leaving England as part of the Task Force, the STD had been given the permanent issue of the latest radio receiver capable of automatically intercepting, scanning and capturing radio frequencies within the ultra high frequency (UHF), VHF and HF wave bands. The STD had carried out field trials on this radio set prior to its permanent issue and was very impressed with the results. This radio set was designed with the then state-of-the-art technology, including in-built computer chip capacity to perform a multitude of tasks such as the intercept of one or more pre-programmed individual frequencies, or all activity within a certain frequency spectrum over one or more bands of frequencies. The set was perhaps if anything a little too sophisticated and versatile to be used to its full capability by the STD, but for the functions we did use it proved invaluable.

From the commencement of establishing an operational state of readiness within the EW centre, this radio receiver was programmed to automatically search for signs of activity within the VHF wave band in the knowledge that the chances of intercepting anything were negligible until the STD was much nearer potential enemy locations. However, there are exceptions to the norm as we were to find out, particularly when radio signals are travelling over water with no obstructions between transmitter and receiver.

At 0819 GMT on 15 May, the automated intercept receiver indicated there was activity within the VHF band. At first, because we were located several hundred miles from the nearest land and not in the path of any known shipping lanes or air corridors in which civilian aircraft would be flying, we assumed this initial intercept to be a spurious transmission. In the event the activity could represent an active transmission of interest and the frequency was passed to WOII Andy Grey, who at the time was carrying out a search programme on a dedicated VHF receiver. Imagine his delight when, tuning in his receiver to the given frequency, he discovered the transmission was in Spanish; furthermore the intercepted activity proved to be military, probably ground-based, in content. Switching on a recording device and simultaneously producing a written translation in English, it was obvious that the transmission was a military network engaged in hostile activities against a British target.

From a quick analysis of this intercept we were aware that the Spanish-speaking operators, who, by vocabulary and other features of recognition had been identified as Argentinian, were losing their battle against a smaller British force. Several Argentinians had been killed or wounded and considerable destruction and damage had been done to equipment, buildings, vehicles and aircraft, in particular Pucara aircraft – the Argentinian Air Force had a number of them in service. Powered by two turboprop engines with a crew of two and for use in a ground attack and counter-insurgency role, the aircraft was fitted with two 20mm cannons, four 7.62 machine guns and three hardpoints for up to 3,300lbs of gun pods, bombs rockets, mines, napalm tanks or torpedoes. Initially I did not know what to make of this intercept; to the best of my knowledge the Task Force was still several hundred miles from the Falkland Islands. We had not disembarked any of HMS

Intrepid's Special Forces but at the same time the STD had not been privy to any Own Force plans to deploy a small attacking force.

On the conclusion of this hostile activity, comprehensive reports of our intercept were sent to our sponsors in England, 3 Brigade HQ on HMS *Fearless* and to the EW coordinating cell under Rear Admiral 'Sandy' Woodward, Task Force Commander on HMS *Hermes*. The only response I received was from the *Hermes'* EW Co-ord Officer Lieutenant Bell RN who wanted to know who I was and how I had come to be in possession of such material. One would have thought that as it was twenty days since the STD was included as part of the Task Force, and at that time the only Army EW unit within the force, the officer in charge of the EW co-ord cell would have been aware of all EW assets at his disposal. The STD at the time of intercepting these communications was, through the absence of direction-finding facilities or third party collateral information, unable to locate the forward Argentinian elements under attack or the Argentinian elements to the rear for whom the transmissions were intended. Later it became clear that the activity we had been reporting on was the Special Forces' attack on Pebble Island. We were unable to intercept the Special Forces' radio communications, who may well have been operating under the cover of radio silence.

Not all ships within the Task Force had been despatched south at the same time, therefore 'catch-up' time was allowed and by 18 May, with hundreds of miles yet to sail, all participating ships were assembled in variably sized flotillas to proceed as a 'fleet' to the Falkland Islands. Several locations and beachheads were rumoured to be the landing site for the ground elements of the Task Force, but the actual objective was not for general dissemination and the Task Force sailed with no knowledge of a final destination, apart from a few headquarters staff. In the absence of information, speculation was rife; one of the most predicted locations was San Carlos. Prior to the landing of the Land Force, Rumour Control speculated that the fleet would sail south past the Falkland Islands, turn 180 degrees to give the impression of heading for Stanley and at the last minute make as if to turn away from the Islands, heading north in the direction from which the fleet had come and giving the impression that the job had been done. We were wrong.

The crew was distracted from too much gossip by a need to complete the cross decking process. On leaving Ascension Island the procedure to ensure that all personnel and equipment was located on the correct vessel had taken more time than allowed for. Consequently, having left Ascension with the task unfinished, the process continued intermittently as the Task Force headed south.

On 18 May a Sea King helicopter transferring 21 men of the Special Air Service (SAS) to HMS *Hermes* was struck by a large bird, possibly an albatross, a fairly common sight flying over the South Atlantic Ocean and with a wing span as large as 3.4 metres, a hazard to low flying aircraft. The helicopter plunged into the Atlantic Ocean, and along with other vessels in the fleet, HMS *Intrepid* was instructed to alter course to search for survivors. Unfortunately the accident happened towards the end of the day when the light was fading and on reaching the area of the accident only two men were rescued. At the time I was not aware of this incident. The following morning I joined two young naval officers at the breakfast table, both looking rather bruised and battered as if they had done ten rounds with Muhammad Ali. I, in my ignorance, passed a comment about their being in a fight; they replied that they were the air crew of the helicopter that had crashed with such a catastrophic loss of life – I wished a hole would swallow me up.

While not personally knowing any of the deceased, much later during the conflict mail addressed to a soldier serving with the Royal Corps of Signals was given to the STD for safekeeping. The addressee was a young man and the son of someone I had served with as an apprentice tradesman at Harrogate in the late 1950s.

Much has been written about this tragic loss of life, and not being an expert on helicopter crashes I am in no way able to offer opinions as to the cause. However, as an observer on the vessel involved in the attempted rescue, I was aware that the only two to survive this unfortunate accident were the air crew, who wore Once Only survival suits. As I was to witness later in the conflict, members of the SAS carry various weapons and huge quantities of ammunition in bandoliers normally slung around their necks and in the pockets of their clothing; the air crew on the other hand were not carrying any additional extra weight.

The Arrowflex machine we used to encrypt reports prior to their transmission by radio proved to be a tremendous asset in the speed of preparation of our reports for distribution to our customers back in England. All activity reports we produced were highly classified and to conform to security regulations had to be encoded prior to transmission. Without the Arrowflex this would have been a time-consuming chore as the content of all messages would have to be prepared off-line by the use of various encoding and decoding devices, with the end product only transmitted over dedicated highly secure lines of communication. The major problem for the STD was that once the report had been encoded, how was it to be transmitted? During our exercise deployments with TCW we only had to pass its reports to the communications centre, from where it would go over dedicated circuits to our customers.

The communications centre on HMS *Intrepid* was, unfortunately, not equipped with similar integral dedicated resources, therefore all material originating to and from the STD had to be transmitted over various secure naval lines of communication. The means to speed up this operation lay in the use of satellite communications but *Intrepid* was not equipped with a 'Scott Box' ship-borne satellite system terminal.

A standard procedure used by the Royal Navy that I, being in the Army, found difficult to accept, was the requirement for the Captain to countersign and approve all forms of written correspondence and documentation leaving his ship. This meant that every message originated by the STD addressed externally from *Intrepid* had to bear the Captain's signature before it was handed into the communications department for onward transmission. Within my own specialist environment it was customary for the originator of a communication to sign for its release unless that person was of a very junior rank, in which case the signature of his immediate supervisor was needed. The only time a written communication would normally require a release signature by a head of department would be if the precedence of the communication was higher than Priority.

By 20 May 1982, the STD was continuously monitoring six Argentinian radio networks believed to be located within the Falkland Islands, plus a few others known to be located on mainland Argentina. Of these by now regularly intercepted voice radio networks, the one considered of greatest importance based on the amount of 'real time' intelligence gleaned from the content of its traffic, was the one we identified as being the 'Malvinos Command Net'. It was established, through passive factors of recognition – the STD had no active factors such as radio direction finding (RDF) equipment at its disposal – that the control of this radio network with its headquarters located in, or within close proximity to, Stanley had regular subscribing outstations located on East Island at Goose Green, Darwin, Fitzroy, San Carlos and on West Island at Port Howard and Fox Bay. Most of these outstations were responsible for controlling subordinate networks of their own. One such subordinate network with its headquarters located at Goose Green and outstations detached to areas around San Carlos and Fanning Head was regularly addressed using call sign '*Guemes*'. Various attempts have been made to interpret/translate this word but its true definition is not known.

From our intercept of this *Guemes* radio network, it was apparent the unit was responsible for patrolling the area north of the bay at San Carlos to Fanning Head (near the entrance to Falkland Sound with clear views stretching out across the South Atlantic). The personnel strength of subunits located in the Fanning Head area according to our intercepted strength returns between the control headquarters in Goose Green and its outstations was approximately 32. This information was of great importance because, as previously mentioned, one of the areas under consideration for a landing by the Land Force was San Carlos and if the defending force was small, the British would initially be virtually unopposed. The information was immediately passed to the Land Force headquarters located on HMS *Fearless* but whether this proved to be part of the final decision to land at San Carlos I was never told.

Apart from the date of 20 May being my wife's birthday, it was also the date on which I wrote what could well have been my last 'bluey' (the Air Mail blank writing paper which when folded was also the envelope). All servicemen when committed for active service and prior to the eve of battle are encouraged to participate in the sombre experience of writing a letter to their next of kin along the lines that it may be the last letter they will ever write. This was to be held unless the author died in action, at which time it would be passed to their next of kin. Writing them proved to be highly emotional created a chastening atmosphere, with the whole detachment sat together in complete silence in the EW cell writing their letters – I was aware of a few sniffles and the wiping away of tears. Overall, not a pleasant experience, especially for the younger members of the Task Force.

8

ACTION STATIONS

Three short blasts from the ship's horn and over the tannoy system would be heard 'Hands to all action stations, hands to all action stations, assume NBC [Nuclear Biological and Chemical] state one condition Zulu.' This was a command very familiar to the STD because it had been activated once or more daily since our departure from Portland dockyard. The broadcast instruction could not be avoided because within the EW cell was a permanently switched-on speaker connected to the ship's command and control internal audio network system.

The immediate action by all personnel on hearing this command was to don anti-flash clothing consisting of a pair of long elbow-length white cotton gloves and a white combined head and neck covering with face mask. The cut-outs for the eyes gave all wearers the appearance of being a member of the Klu Klux Klan. The clothing was to give limited protection to exposed skin in the event of being in close proximity to fires and explosions. Respirators and full NBC clothing were to be on hand, the ship was 'battened' down and the ship's crew had to report to their duty stations. Fortunately, from leaving England on 26 April until 20 May, on all occasions this 'hands to action station' command was given only for training purposes or, as we gradually got closer to the Falkland Islands, the command later turned out to be a false alarm. After 20 May, when we heard this command it was for real.

As previously mentioned, the STD occupied what had previously been the ship's EW cell situated on the deck below the bridge, fairly close to the Captain's day cabin, alongside a couple of cabins occupied by other 'sensitive' units/organisations and in close proximity to the largest enclosed open space on HMS *Intrepid* used by the alternate Operational Command and Control Land Forces Headquarters, duplicating, where possible, the main headquarters established on HMS *Fearless*. It was standard procedure for land forces to establish main and alternate control headquarters as this allowed for each headquarters to maintain a 24-hour effectiveness by 'leap-frogging' each other and in the event of one headquarters becoming non operational, the other, being a mirror image and 'up to speed', could continue operations.

By the time the flotilla carrying the Land Force had reached Falkland waters, secure communications between main and alternative headquarters had been well established allowing all parties concerned to 'sing from the same hymn book'. These secure communications also included a dedicated link between us and the intelligence cell located on *Fearless* in support of 3 Brigade. However, because our dedicated link to Headquarters 3 Brigade was via a VHF voice channel, there still remained the question as to how our operational traffic, encrypted through the Arrowflex machine, could be transmitted more speedily to our sponsors 8,000 miles away in England.

Eventually this problem was resolved by using the P&O North Sea ferry-boat MV *Norland's* satellite communications. The *Norland* was taken out of trade to be used as a troop carrier to transport 2nd Battalion, the Parachute Regiment to the South Atlantic. As the MV *Norland* was not equipped to be a warship, shortly after leaving Ascension Island she sailed in close proximity to HMS *Intrepid* who, in the event of any unforeseen problems, could offer help. This 'togetherness' allowed the communications cell on *Intrepid* to establish an insecure radio link to *Norland* along which was transmitted the content of our fives unit perforated tapes; the tapes were then transferred to a satellite communications system and forwarded to England. This arrangement meant almost instant contact between the STD and home.

Amongst the sensitive units/organisations occupying office space on the same deck as the STD was a WO II Bird serving with the Small Arms Corp. Bird was a 'one man band' who, prior to hostilities had been working on field operation trials of various infra red and night vision equipment. One such piece of equipment he had brought to war was an infra red type of recording video camera. Over our short acquaintance I had established a good relationship with Bird, who allowed me access to some of the footage taken by the trial's video camera and we frequently discussed how he could use the camera to its best advantage if hostilities started.

On the afternoon of 20 May, it was now common knowledge that the Task Force was to establish a beachhead in the area of San Carlos. I mentioned to WO II Bird that I had information from my own source that there was a possibility that only somewhere in the region of 30 Argentinian troops were active in an area close to Fanning Head and that it was a pity he could not be deployed with his night vision capability video camera to confirm this. He agreed with me and said he would see what could be done.

In the early evening and already very dark outside because it was winter in the southern hemisphere, Bird asked to see me in his operational cell. When I entered he was looking at a video monitor displaying a moving picture of what, to me, initially looked like a dark grey background with several small off-white blobs which, because I went for the obvious, appeared to be sheep. Some of the blobs were moving and others were still. He assured me that what I was observing was in fact a group of about 25 people at Fanning Head. After I had spoken with Bird earlier in the day, he had managed to persuade the hierarchy to allow him the use of a helicopter with a volunteer pilot fully aware of the possible outcome of night flying over hostile territory, to assist him in a reconnaissance flight in the area of Fanning Head. Between them they assembled the camera Heath-Robinson style beneath the undercarriage of the helicopter so that when in flight, with Bird hanging out of the open door he was able to operate the camera to produce the pictures I had seen.

Late afternoon of 20 May, with the knowledge that all peace talks had ended with neither side conceding defeat, it was now inevitable that a period of hostile military engagement between the Armed Forces of the United Kingdom and the Argentine Republic would commence at or around first light the following day. It had been agreed between me and the intelligence cell of 3 Brigade, that to avoid a loss in continuity the STD should remain in its current operational role on board HMS *Intrepid*. Meanwhile the Marine Y Tp would disembark from HMS *Fearless*, establish its mobile operational role and, when in a position to commence operations, we would refit its specialist equipment into the Detachment vehicles and join the Marine Y Tp on land. It was believed that barring delays the STD and Y Tp would meet up at lunchtime on 21 May.

By early evening all disembarkation preparations were complete and the crew and forces on *Intrepid* were playing the waiting game. There was a strange feeling of anticlimax. Having travelled so far in a state of constant preparedness and not fired a single shot in anger, now that we had finally arrived at our destination, all those I engaged in conversation were slowly coming to the conclusion that perhaps nothing bad was going to happen. Accepting the night would most likely be long and no doubt hectic, I retired to my cabin in an attempt to at least get a couple of hours of sleep before the action, if any, started. But sleep was impossible with the thought that within a few hours we would move from preparedness for war to participation in an armed conflict.

At all levels within the Argentinian operational radio communication networks intercepted by the STD – in keeping with most communications networks the world over – in comparison with the prolonged periods of activity during the day, they were seldom active during the night, apart from a 'watching brief', where operators would occasionally make contact to ensure their equipment was still working and operators had not fallen asleep. The early part of the night 20/21 May 1982 was to prove no exception.

In close proximity to land and physically that much nearer to the Argentine Force locations, the quality and audibility of intercepts of their radio transmissions had vastly improved and we were maintaining continuity on a regular basis on several radio communication networks, all utilising non-encrypted clear voice or Morse. Most of these networks were the communications between the military headquarters in Buenos Aires, mainland Argentina and General Mario Menendez's headquarters of the Argentinian Ground Forces Malvinos (Falkland Islands). Others were Argentinian naval and a few were communications between the Argentinian Ground Forces located on East and West Falkland Islands. Because these networks were in the HF band and were insecure in as much as there were not protected by on-line encryption devices, so long as the strength and quality of the radio signal was good, anyone with a half decent radio receiver located on either side of the South Atlantic would also have been privy to these transmissions.

One of these networks, as previously mentioned, was the Command and Control Ground Forces with its control in Stanley. This network was number one in our list of priorities and was continuously monitored by a dedicated intercept operator. The geographical locations of the outstations on this network were established through a process of elimination and to references in intercepted material and not by RDF. The only British Army unit with mobile, ground-based RDF capability was under command of 1st

British Corps, BAOR and at the time, no spare equipment was available for emergencies like Operation *Corporate*.

While it is relatively easy to establish locations of outstations on any particular network, it is appreciably more difficult to establish their identities and roles. However, it was believed that located on West Island were elements from 5th Infantry Regiment ('Reconquest Task Force') and 9th Motorised Engineer Company, garrisoned in and around Port Howard, while the 8th (General Bernardo O'Higgins) Infantry Regiment and 9th Motorised Engineer Company, were garrisoned farther south in the area of Fox Bay. The overall commander of all Argentinian Forces on West Island was believed to have been a Brigadier General Parala of 3 Brigade, with Colonel Juan R. Mabragana commanding the garrison at Port Howard, and Lieutenant Colonel Ernesto A. Repossi commanding the garrison at Fox Bay. Identifying formations and units of the Argentinian Armed Forces on East Island was more difficult; apart from the small force of *Guemes* (C Company 25th Infantry Regiment), located in the San Carlos/Fanning Head area, the only other confirmed STD unit identities were 2nd Airborne Infantry Regiment, 12th Infantry Regiment and elements of the 602nd Special Forces, all garrisoned in the Goose Green, Darwin area.

H Hour had been set for 0230 GMT (Falkland Island local time is four hours behind GMT) by which time HMS *Intrepid* and other vessels in the Task Force were in positions at anchor within the Falkland Sound with the entire ship's crew and forces now in a full state of readiness. *Intrepid's* four Landing Craft Utility (LCUs) were despatched to transfer elements of the Paras on board MV *Norland*, with the STD sat waiting patiently for activity to erupt from our radio sets. The first intercepted Argentinian radio activity of the day was believed to be from the area of Port Howard, West Island, outstation to the controlling authority in Stanley on the Command Net, reporting that at approximately 1115 hours GMT a detached element had reported the sighting of British naval activity in close proximity to San Carlos Bay and the vessels seen were a 'huge white vessel, not a warship', 'four large camouflaged landing craft much bigger than normal landing craft', 'a large war ship' and a 'smaller red craft'. As previously mentioned, within the EW cabin of *Intrepid* was a direct secure radio link to the intelligence cell on HMS *Fearless*; I immediately contacted the 3 Brigade intelligence cell to inform them that the activities of the Task Force were being observed by the Argentinians. After about ten minutes I received a call back asking me how I knew the Argentinians had observed the Task Force. Trying to keep calm and composed I replied that I would read the transcribed content of the message passed from the area of Port Howard to Stanley. All I got in return was an 'Oh', but I could imagine from the response the puzzled look on the operator's face, so I then explained the content of the message: the huge white vessel was the *Canberra*; the four camouflaged landing craft were the *Intrepid* or *Fearless* landing craft; the large warship was the *Intrepid* or the *Fearless*; and the smaller red craft was the *Norland*.

Apart from the activity on Pebble Island, this was the STD's first live intelligence report direct to the intelligence cell of 3 Brigade and if their response was anything to go by, it did not bode well for the future. This Argentinian command net activity was followed shortly after by the activation of a radio link serving the *Guemes* in the Fanning Head area and their control authority in the San Carlos area. Content of the intercept on this link related to the fact that the *Guemes* had come under fire from a small attacking force.

I soon saw my first war casualty. A young black soldier or marine, with what appeared to be severe head and facial wounds, lay on a stretcher in a gangway while a group of sailors with metal cutting equipment were attempting to make an entrance way wide enough to allow the stretcher to be carried through and into a small make-do emergency medical facility in the gun room. This was located behind the bar of the ward room and had previously been used to store kegs of beer. My enquiries as to how the man on the stretcher received his injuries revealed that it was most likely a 'Blue on Blue' situation as forward patrols of the Paras and Marines crossed each other's path, neither group having been briefed to expect to meet friendly forces. Obviously, little attention was paid to my intelligence indicating the immediate opposing force consisted of about 30 *Guemes* who had been dealt with much earlier in the morning.

The disembarkation of the ground forces proceeded virtually unhindered, and through the thorough professionalism of those involved, a bridgehead within San Carlos Bay was quickly established. At a little after midday, with the morning sea mist lifting and first light penetrating the cloudy sky, members of the British Task Force were to experience their first Argentinian air activity when, as if from nowhere, two Argentinian Air Force Pucara aircraft – believed to be based at Goose Green – flew over the San Carlos Bay anchorage on what was thought to be an initial reconnaissance flight. This possible Argentinian reconnaissance sortie was at approximately 1245 GMT on 21 May and was followed by various aircraft from mainland Argentina who proceeded to fire missiles at and drop bombs on the Land Task Force.

HMS *Intrepid's* weaponry consisted of the Seacat GWS-20 missile system, 40mm Bofors guns and nine-barrelled launchers for throwing chaff. Whenever the missile systems were activated or the ship's weapons brought into use, orders passed to the weapons crews were heard through the communication speaker located in the cabin. One of the 40mm Bofors guns was situated on the ship's bridge wing directly above the EW cabin. When this gun was fired those in the cabin knew exactly how many shells had been used because when the gun was reloaded the ejected spent case from the breech crashed to the floor above our heads; at times we thought the case was actually coming in through the ceiling.

The Seacat missile system with quad launchers, as fitted to HMS *Intrepid*, is a short-range surface to air missile, brought into service with the Royal Navy in 1962, originally to replace the 40mm Bofors gun. Targets are acquired visually with the missile guided via a radio link by the operator inputting commands through a joystick. After firing, an interval of about seven seconds elapses before the missile is gathered into the operator's sight and brought under his control, which means the system is not effective at ranges closer than 500 metres.

During the evening meal in the ward room on 21 May, the main topic of discussion was how lucky *Intrepid* had been not to have received any direct hits from the Argentine Air Force, even though several hundreds of pounds of explosives had landed in close proximity. It also came to light that the ship was even luckier not to have been hit by a Royal Navy Seacat missile which, during a return of fire, passed directly over her bridge. According to Rumour Control and the 'kangaroo court' established by crew members, a missile had been launched by another Royal Navy ship also at anchor in San Carlos Bay, and because all ships fitted with the Seacat were issued the same crystal frequency settings prior to departing the UK, more than one ship was using the same frequency for the radio link to the missile guidance system.

As mentioned, once the missile has been launched, it takes almost seven seconds before the missile is under the operator's control. Therefore we had a scenario where it was believed two missiles were launched a matter of seconds apart. After approximately seven seconds the first missile was locked into the operator's guidance control system through the joystick, while seconds later another missile was fired from a different launcher. But because the radio frequency to guide both missiles was identical, seven seconds after launch of the second missile, it locked into the already operational guidance system of the first missile launched. Because the target is acquired visually and the missile guided likewise, the operator of the first missile launched would have no idea that he was also guiding a second missile. Which naval ship came so close to hitting HMS *Intrepid* at the commencement of hostilities will never be known but her actions quickly ensured a firm control over crystal frequency usage by all ships fitted with the Seacat.

The first Argentinian Air Force strike had delayed and thrown into disarray the STD's plan of action for that day. Consequently the time for us to meet on land with the Marine Y Tp had long since passed without any instructions for the STD to prepare to disembark. Much later in the day I was informed that it would be better for us to remain operational on *Intrepid* for the foreseeable future because when the Marine Y Tp disembarked from HMS *Fearless*, their Land Rovers had become bogged down in the soft sand on the beach and would need to be recovered. As theirs was a low priority it would be quite some time before recovery arrived. Also, to the front of where the vehicles were stuck was an almost vertical and very high range of hills, and although Y Tp were in possession of all current enemy intercept target details, they were unable to intercept target activity because the aerials were below the level of the high ground, screened from the majority of Argentinian radio signal footprints over East Falklands. On the other hand, because *Intrepid* was some distance from the high ground, and by using the ship's aerials which gave us added height, the STD intercept of the target array was not affected.

The next time I met with members of the Marine Y Tp was in Stanley about two weeks after the cessation of hostilities. It was apparent that the EW element of the Y Tp had been totally non effective for the duration of hostilities. I personally did not see a single enemy activity report that could have been attributed to Y Tp resources throughout the war. It transpired that with the lack of suitable roads or tracks their Land Rover-fitted operation platform could not be moved, and when the order for the Marines to 'Yomp' to Stanley was given, the radio equipment added to all personal equipment was far too heavy to carry. Their only means of moving forward with the advancing troops or to a forward location suitable for a deployed Sigint site was by helicopter, but of course none was available. I was aware that elements of the Y Tp were effectively employed within the 3 Brigade intelligence cell and later in the Land Task Force intelligence cell.

As a commissioned officer, during previous tours of duty one of my many secondary duties had been that of Communication Security Officer

(COMSO), not an onerous duty but one that carried immense responsibility. As OC the STD and the only officer in the unit, the duties of the COMSO once again fell to me. The COMSO was ultimately responsible for the security, handling and storage of all classified documents, signal traffic and encryption systems and for ensuring that personnel handling classified material in the pursuance of their duties had the necessary clearances.

During one of the many briefings I attended prior to leaving the United Kingdom, I was informed that no matter what else might happen, for security purposes the operational mode, standard practices and procedures and our product would always be in adherence with 'peacetime restraints', as contained in various government manuals and routine instructions relating to Sigint. Contained within these security manuals were instructions for the storage of classified material which, as soon as I set foot on HMS *Intrepid*, were contravened because the accepted secure storage containers were not available to store the data and material given to me by the organisation that placed the restraints on me in the first place.

On the way south and finally recognising the fact that hostilities were inevitable, I decided that because of the possibility of capture of the STD by the enemy, I had better consider taking precautions to reduce the amount of classified and very sensitive material we held. Because of its classification, the only acceptable method of destruction of the materials in my possession was by either cross-cut shredding with a minute separation between the cutting blades or destruction by burning. *Intrepid* was not equipped with an incinerator or a cross-cut shredder conforming to the required standards, so the only means of destruction was the Navy way, which was by means of a large canvas pouch, lead lined for increased weight and lockable with steel bars, hasp and staples and padlocks. The items for destruction were placed inside the pouch, padlocked and dropped over the side of the ship. The majority of Sigint content material retained for archives is either kept indefinitely, or for a stated period of years, before declassification or destruction; either way it will be eventually removed and destroyed. I believe that the Navy's method of disposal, namely long term storage at the bottom of the South Atlantic Ocean met all the 'peacetime restraints' requirements. With the majority of classified material now lying at the bottom of the ocean the limited amount left was stored in what would have started life as an iron wall-mounted safe complete with secure combination lock and used by previous occupants of the EW cell. After this initial destruction of classified material any new material retained would have to be of such importance that we couldn't possibly perform our operational role without it.

The destruction of classified waste over the years has inevitably been the cause of many problems. The first time I encountered it was during my first tour of duty in Cyprus. The unit there generated mountains of paper waste each week, all of which had to be burnt. At the time the incinerator used for this destruction could best be described as an 'ox spit'. The incinerator had a central spindle supported on four legs; a handle was attached to one end of the spindle to rotate a very large drum-like cage made of iron mesh. The drum had a door through which the waste was placed, and after a good soaking with petrol the contents were set alight. The turning of the drum accelerated the burning but it also sent partially burnt paper flying from the incinerator out of the secure area and onto waste land adjacent.

This method of classified waste disposal continued for a couple of years, until one day a local shepherd who grazed his sheep and goats on the wasteland near to the unit turned up at the door of the operation building demanding to speak to the officer in charge. He was introduced to the Senior Intelligence Officer and informed him that his animals were frequently eating part-burned pieces of paper. The shepherd then presented the officer with a bag of charred paper. Closer inspection revealed perfectly legible parts of highly classified and sensitive information. The solution was to build several drums inside each other with the diameter of the mesh getting progressively smaller, so that paper escaping through the first drum would not get through the outer ones.

The problem with the disposal of classified waste at this unit in Cyprus was highlighted in what became known as the Cyprus Spy Trial of 1984, when a number of servicemen serving at the unit were accused of spying for the Russians. Although at the time of this particular incident I was not serving with the unit and therefore have no actual proof, only the words of some who were, I gather one of the principal offences committed that prompted prosecution involved the illegal removal of classified waste awaiting destruction. The destruction of classified waste was carried out at various times during the day and night and entirely dependent on the availability of individuals. So a classified document has been properly stored and accounted for but has reached its 'sell by date' – what would normally have happened to it? The document would have been removed from its secure storage and deleted from the classified documents register – this procedure would have been carried out and witnessed by individuals who had the necessary authority – the document would then be placed in a large burnable brown paper sack and securely stored with other material for incineration when personnel were available to do so. Then the paper sack would be removed from the secure store and taken to the incinerator.

The weak points in the chain were when documents were in transit between the person authorising the destruction and the secure store, and/ or the secure store and the incinerator, at which points the paper sacks were left open at one end and not securely closed and sealed. Therefore, all those involved in the sack handling chain would have had visual sightings of the contents at the top of the brown paper sacks with the opportunity of removing some of the documents and discreetly hiding them about their person, safe in the knowledge that others believed the documents to have been destroyed in the required manner. As you can imagine, once this now obvious omission of sealing the top of the sacks came to light all orders relating to the handling of classified waste were very hastily amended.

HMS *Intrepid* had a variety of operational roles during Operation *Corporate* and most of them were carried out after dark. The Special Services were amongst the embarked force and they had established a controlling function within the area of the operations room from where they would communicate with patrols on both islands. These penetration patrols would require replacement and replenishment if they were to carry out their tasks effectively and it fell to the crew of *Intrepid* to provide the transportation to allow them to hop around the islands. At a prearranged time after dark, *Intrepid* would leave its moorings, enter the Falkland Sound, steer to the left through the channel between East and West Island, then onward to the designated locations.

Intelligence reports suggested that somewhere in close proximity to Fox Bay, the Argentinian ground forces had established a radar site with clear vision across the sound and open waters of the South Atlantic. In addition to the radar site there was also a medium artillery unit located close by; presumably once a potential target had been identified on the radar screen, guidance would be passed to the artillery. The STD's role during these night operations was to monitor for active communication between all known units, particularly those loosely connected to artillery and radar assets, within the immediate area of Fox Bay and to act as an early warning for *Intrepid's* approach through the canal to open water. A further safeguard was a Royal Navy frigate, on station to provide covering fire power as *Intrepid* passed the nearest point to the radar site.

On completion of the night's mission, *Intrepid* would return to its anchorage in San Carlos Bay with the same protective measures taken for the return journey as those taken for the outward one. Irrespective of the night's activities, *Intrepid* was always at its anchorage by first light ready to receive the first of the Argentinian bombs for that day.

After the cessation of hostilities when the STD had temporary operational accommodation in Government House, I was given a Spanish language photocopy of *La Gaceta Argentina*, a report compiled by Senior Lieutenant Carlos Daniel Esteban, OC C Company 25th Infantry Regiment, on the part played by the *Guemes* at the time of the landing of the British Land Forces on the Falkland Islands. An English translation of this article is included as Appendix 1. I will leave it to the reader to determine how accurate the facts reported by the STD were in its real-time intelligence reporting of the same event. However, it would appear from his report that the *Guemes* sighting of the arrival of the Task Force was some two hours after the initial report from Port Howard to Stanley.

9

LIFE IN BOMB ALLEY

The assault ship HMS *Intrepid*, capable of landing up to 750 troops on shore from landing craft and helicopter was thought to have come to the end of her useful life in the British Navy and had been mothballed in preparation for her eventual disposal. Operation *Corporate* put paid to that and she was hurriedly prepared for war.

As a result of the late decision to include her and the time taken to fully prepare her for fitness, most of the Task Force had left British shores and was well on the way to Ascension Island before her departure from Portland dockyard, which was considerably quieter than that given to other departing ships and although not intentional, it is possible that apart from her crew and a limited number of embarked troops, not many others were aware she had sailed to join the Task Force. Whatever the reason for her low profile it was apparent that the Argentinians did not know for some considerable time that *Intrepid* was included as a member of the fleet and therefore the ship was not initially included in the Argentinian listings of the British order of battle.

There was a possibility that *Intrepid*, because of her position while at anchorage in San Carlos Bay, had avoided detection by being in the shadow of enemy aerial reconnaissance. So, in order not to advertise the fact that she was operational in the South Atlantic and that she had survived the initial Argentinian bombing unscathed, it was decided that by day she should continue to occupy the same anchorage, later to be referred to locally as 'Bomb Alley'. For nearly four weeks during daylight hours, Bomb Alley was to be

the home location for *Intrepid* and the STD and consequently a target for all
that the Argentinian Air Force could throw at them.

After the first day's assaults on the Task Force in Bomb Alley, it was obvious to me that the lack of ground-based radar to intercept the Argentinian
aircraft meant that those vessels at anchorage in San Carlos Bay and the
ground forces in and around the bridgehead were sitting ducks for aerial
bombardments. The majority of the Argentine Air Force missions would
depart from an airfield in Comodoro Rivadavia in the south of Argentina
and some 600 miles from the Falkland Islands, and with a bit of luck, if their
targets were on or in the vicinity of the Falkland Islands, their approach
would be detected on the radar screens of a British warship on point duty
to the west of the Falkland Islands, and the headquarters of 3 Brigade would
be immediately notified of any imminent danger.

However, what the British headquarters did not know was the estimated
time to target and the direction of approach. Therefore, before the troops
on the ground had time to take preventative action, the enemy aircraft were
above them dropping bombs. After a couple of days suffering at the hands
of the enemy air force, the STD were aware that the Argentinian aircraft
departed from their mainland airfields under cover of radio silence and on
reaching the Falkland Islands would break it for the pack leader to issue
instructions to his subordinate pilots to approach the target area, over-fly
the bay, isolate a specific target for their return run, drop their bombs and
return to base. The STD had in its database the parameters of radio frequencies of all the aircraft used by the Argentinian Air Force. These frequencies
were entered into the memory of our new automated receiver, which then
locked onto aircraft radio frequencies immediately they were activated,
confirming that if the enemy aircraft radios were switched to a transmit
receive mode, then the aircraft were very likely in close proximity to the
Falkland Islands and possibly an attack was about to take place. It was agreed
with the Captain of *Intrepid* that the simplest and quickest means available
for the STD to alert the Task Force to incoming enemy aircraft was to
inform the Captain of *Intrepid* who would immediately signal via the ship's
horn an impending attack on vessels in the anchorage and those within
close proximity to the beachhead.

In addition to this limited and unsophisticated form of early warning
set-up was the intercepted material gleaned from the command control
network outstation at Port Howard. A small detachment from Port Howard
had established a forward observation post in the vicinity of Many Branch

House, located north of Port Howard, West Island, and it is believed it was from here that the British landing in San Carlos Bay was first observed and reported back to Argentinian High Command. This forward observation post was at times used to direct Argentinian aircraft to potential targets within San Carlos Bay and therefore, when activated, gave the STD limited additional warning of approaching enemy aircraft over the bay.

After about four days of reporting on the activity of this outstation, I was informed that an SAS patrol would be inserted into West Island to take out this observation post. I strongly opposed this proposal on the grounds that although our attempts to contribute towards early warning of incoming enemy aircraft were limited, the intelligence gained from this outstation on West Island was better than none, as the Task Force had no other early warning facility. Obviously my request not to disturb this enemy observation post was ignored because not long after I heard of the death of SAS officer Captain Hamilton MC, apparently during his section's attempt to remove the observation post, or at least observe it.

On a previous tour of duty in BAOR in the mid 1970s as the OC of an Elint mobile operations troop (believed to have been the only such unit of its type within NATO at the time), I was responsible for conducting field trials as to the vulnerability to detection of the Rapier missile system through the intercept of its various radar components by a potentially hostile organisation. The trials were initially carried out on the weapon system for use in the defence of strategic locations such as airfields and other large strategic installations, then as a tactical weapon on the battlefield. During 1 BR Corps level and NATO Field Training Exercises (FTX) held in West Germany, the Rapier units were constantly exercised in their specialist role of providing air defence for Corps troops. They would practise both carpet defence – deployed to give protection in depth and width – and route defence to provide protection for columns on the move. The fact that my little mobile Elint troop could intercept and locate with ease the Rapier weapon system when deployed on an FTX was the fault of the lack of specialist security in the design of some inbuilt components and in no way detracted from the professionalism associated with the Royal Artillery. As a result of my practical experience and knowledge of the Rapier system I failed to accept the logic of any armed force contemplating an action the size and role of Operation *Corporate* without the inclusion of at least a battery of Rapier missiles to be landed with the first assault troops in order to provide essential aerial protection. The Task Force did have Rapier systems at its disposal but most of these were in the hold of the *Atlantic Conveyor* when it was sunk in the South Atlantic; others were with the Task Force but the unit's disembarkation was of such a low priority that by the time the missile battery was unloaded, the Argentinian aerial bombardment was well underway and had inflicted considerable damage.

The Rapier missile system in its mobile form was designed for the entire system to be carried in a converted RCM748 armoured track vehicle. Those landed on the Falkland Islands were Land Rover-towed, wheeled versions. The terrain around San Carlos Bay precluded the use of wheeled vehicles; therefore the Rapier system had to be transported by helicopter. In the event of a landing on the Falkland Islands, the Rapier's deployment locations had previously been chosen by computer back in the UK. The locations were designed to protect the Task Force landing itself, not units of the fleet stationary in Falkland Water some distance from the shore. On their eventual deployment the Rapier installations were confronted with some unusual targets; because so many Argentine attacks were directed at the ships of the Task Force rather than its land units, the Rapier operators found themselves aiming the missiles downwards into mist-shrouded valleys and not upwards into clear skies. I have often wondered since what would any Commander of 1st British Corps have said to his staff if on exercise deployment his aerial defence system had been left behind in barracks and his troops had been deployed without air cover confronted with what on paper could have been a superior – if only in numbers – air force than his own?

The control station of the dedicated radio network serving the Special Forces was located within the operations centre of *Intrepid* and in common with most British Ground Force radio operators, the operators manning this control station could not comprehend the fact that 'less is best'. All Clansman radio sets in use with the British Services in the 1980s were manufactured with a manual control device to increase or decrease the strength of signal output from the radio set, but most operators assumed the only way these control devices would work was to have them set at maximum. The HF band is an extremely busy environment in which to communicate with all users vying for the best frequencies, which was no doubt why the frequency used by the Special Forces' controlling net was very close to that used by the Argentinian Ground Forces Command and Control Network, the STD's number one intercept priority target. The effect of this closeness was to invariably 'jam' each others' transmissions making it extremely difficult for us to maintain continuity on the enemy transmissions. Initially, when the 'jamming' of the Argentinian network was obviously caused by the Special Forces' control net, a member of the STD would engage the operator of the latter network in conversation, then without making it too obvious, reach out and turn the output volume control down. This devious method would only last a couple of hours and then the operator would increase the output to maximum.

After a few days, the Argentinian operators had had enough of being unable to communicate through the interference caused by the Special Forces communications, not that they knew who it was, and requested the perpetrators be located through the Argentinian mainland RDF unit. Not knowing how efficient Argentinian RDF was, and not prepared to find out, an official request was made for the Special Forces radio network to operate on very low power; after all, most subscribers to the network were located in a 10–20 mile radius of the control facility. This request had its desired effect; the radio sets' output was turned down and kept down, the Argentinian network no longer complained about the 'jamming' and we heard no more about Argentinian RDF.

Listening to Argentinian radio conversations between the outstation located at Port Howard and its control in Stanley, it was obvious that the penetration patrols of the SAS into the area of Port Howard were on occasion observed, generally from the observation post located at Many Branch House. After the first occasion of an SAS penetration patrol being observed I decided to inform the OC SAS Forces embarked on HMS *Intrepid*. I believe this person was OC B Coy. When I first asked him if he had subordinate troops on West Island he flatly denied any knowledge but then asked me who I was and what organisation I represented. When I informed him his manner changed and he admitted that at times SAS patrols were active on West Island. After our first meeting it was agreed the STD could be of some help to his organisation by alerting the radio network control operator whenever a penetration patrol was observed. The control operator would then either inform the patrol to go to ground or to withdraw from the area.

When stationed in BAOR, I had frequently participated in 1st British Corps exercises employed either in the EW cell or as a member of the enemy ground forces input team. For both types of specialist employment I would be located within the field intelligence cell at the corps HQ along with other specialist organisations/units including representatives from the

SAS. Within the intelligence cell, each sub unit would bring a degree of experience and expert knowledge to assist the intelligence staff, in particular the G1 Int, to conduct the exercise in as realistic and authentic a way as possible. All very large scale training exercises would be run from the 'Pink', the written instructions as to who will do what, where and when. The Pink would be written to create a realistic scenario but also allowing for minimal extra input as the exercise progressed.

In addition to the Pink was the 'Bird Table', a very large, three-dimensional, small scale papier-mâché model of the terrain over which the exercise was to be conducted. On the Bird Table would be various scaled models and locations of 'blue' (own) and 'red' (enemy) forces. Movement on the Bird Table was officially carried out by the intelligence cell, and was continuous but always agreed with the written word as contained in the Pink. During one such 1st British Corps FTX, I was assisting in playing the role of the 'red' force, elements of the Group of Soviet Forces East Germany. The location of the exercise was the BAOR area of responsibility within West Germany, with the forward edge of the battle area being the East/West German border. On the second day's play I placed on the Bird Table a model of a Soviet helicopter but my action was never challenged. Four days later I informed the G1 Int (Lieutenant Colonel commanding the intelligence cell at Corps HQ) that the unopposed helicopter landing alongside a lake in the Hartz Mountains carried troops of the Soviet Special Forces who had proceeded to contaminate the lake with poison. The four-day delay in bringing this to the attention of the G1 Int was to allow the water within the lake to percolate down and enter the main water supply used by those in the geographic area covered by the Bird Table. It was customary for troops when on an FTX to refill water containers from local garages within the immediate area. The only response I got to my initiative was 'the exercise cost a lot of money to organise and control, therefore it must run the full duration.' Participation in these corps exercises was the closest the specialist worlds of Sigint and the SAS had got without there being a direct exchange of each other's product. Now, during Operation *Corporate*, without any go-between, both organisations were in direct dialogue with each other.

Approximately a week after establishing the bridgehead at San Carlos, the Argentinian forces at Port Howard were aware of a British SAS patrol active within their area of responsibility and patrols had been deployed to search for them. Several sightings had been made by the Argentinians of this SAS patrol and on each occasion the control operator of the SAS radio network on HMS *Intrepid* had been notified of our findings. Later in the day Port Howard reported to its control in Stanley that after a fierce and bloody encounter with an enemy patrol of two men, one had been killed and the other wounded and taken prisoner. Included in this intercepted message were the personal details of the two men involved. I immediately passed this information to OC B Coy who then asked me which of the two SAS personnel had been killed and who was taken prisoner because from the intercept this was not apparent. Included in the personal details passed to Stanley was the date of birth of a captain, but not for the corporal; therefore, I assumed, wrongly as it happened, that the captain had been able to give his own date of birth but would not necessarily have known the date of birth of the corporal. I discovered later that the deceased was Captain Gavin John Hamilton originally of the Green Howards.

No matter what prior preparation and action is taken nothing can pre-pare an individual for the way he or she will react to coming under aerial bombardment for the first time. As a very young child I was raised in the West End of London for most of the Second World War, and throughout my early life I recall being woken from my sleep almost nightly to be taken by my mother to an air raid shelter. Although we had an Anderson air shel-ter in our back garden my mother preferred to take us to one of the larger neighbourhood communal shelters; maybe she thought there was safety in numbers. Having spent the night in the neighbourhood communal shelter lying on a makeshift bed and listening to bombs exploding all around us, in the morning on leaving the security of the shelter, I would often have to climb through the wreckage of buildings demolished during the night. This became a way of a life and one that I am pleased to say with the passing of time is almost forgotten. The rest of the STD team were too young to have experienced the horrors of live aerial bombing before they entered Bomb Alley on 21 May 1982.

As a soldier fighting a land battle you learn that one way of attempt-ing to escape enemy bombing is to dig a hole, and as the days go by and the bombing continues the hole gets deeper and deeper. Unfortunately, on board a warship this is not possible and unless your specialist employment is in the defence of the ship and its crew, there is little else to do on hearing the air raid warning other than to 'hit the deck'.

'Hitting the deck' is simply just that, from the vertical you adopt the horizontal position as fast as possible. The reason for this, we were reliably informed, was because if a ship were to take a direct hit, on the explosion of the bomb or shell various shock waves are sent out. Anyone in the upright position hit by the force from these waves could suffer broken bones; if on the other hand the person is lying horizontally on the deck, most waves will pass through the body as they pass over the decks.

On hearing the command 'hit the deck' for real, those members of the STD on duty in the EW cell at the time, couldn't have fallen any faster, but then we stumbled on a problem. Because the EW cabin was very small, there was insufficient room for more than three people to lie on the floor. As the person in charge I was responsible for the safety of my team and therefore the last person to hit the deck having made sure all the team were in position first. However, at this stage with all the floor space taken, all I could do was to ask the team to squeeze up to make room for me and sufficient space was then made to allow me to sit on the floor with my legs outstretched but with my back against the iron security safe. While in this position, with my head exposed above the top of the safe I constantly thought that during an air raid, if the ship was to be hit by an Exocet missile, I would be decapitated.

During an air raid a strange kind of peace and serenity would descend on the ship; conversation and movement stopped as the majority of the crew and embarked force lay on the decks in complete silence. That silence was only broken by the sound of aircraft flying fast and low over the target area, the crashing of the empty shells landing on decks above when the anti-aircraft gunners ejected the cases to reload, the sound of bombs exploding close by and the command and control instructions issued by the *Intrepid's* Captain to the crew through the internal audio control network.

Listening to orders and instructions from the internal control network I could not but be impressed by the way *Intrepid's* Captain and Commander controlled the ship's duty station crews during attacks by aircraft. Not once did I detect any indication of anxiety in their voices and their calm and collected manner helped to put those at ease who showed obvious signs of distress and agitation. The 'all clear' signal was usually followed by 'one all round' and while the smokers were busy polluting the atmosphere, the topics of discussion were near misses and the pleasure of having survived the onslaught, and an air of normality would return to the ship.

Within the EW cell, after the first few air raids it was considered that when in the 'hit the deck' position too much vital information could be lost. By extending the length of the leads connecting the headphones to the radio sets, continuity could be maintained when lying on the floor, then on the 'all clear' it was back to the chairs and working on the desk tops. Within minutes of returning to normal working conditions I became aware that when Jock, an Intelligence Corps corporal, was on duty, he would ask to leave as he wished to make a visit to the 'heads'. After several days of the same routine I casually mentioned that he must have a small bladder if he couldn't wait until settled in properly after an attack. He replied that it wasn't his bladder but his bowels that he was concerned about; during raids he became so frightened that at times it became very difficult for him to stop his bowels opening.

Most soccer fans will be aware that 1982 was a World Cup year, with the finals played in Spain and Italy the eventual winners of the competition. England and Argentina both made it to the final rounds but were knocked out before the possibility of having to play each other. Apart from the occasional mention on the BBC's World Service, the English supporters in the South Atlantic were denied knowledge of what was happening during the build-up to the final unless they spoke Spanish and were able to tune their radios into commercial radio stations in South America, but from my observation there were not many Spanish linguists serving with the Task Force.

At times the workload in the STD reached saturation point with delays in translation caused by the lack of competent linguists. Realising the situation, *Intrepid's* Captain offered me the use of one his junior officers; unfortunately the officer's language skills were not up to real time translation and his services could not be used. The STD's sponsors in England were aware of this shortage of expertise and I was informed that six RAF personnel had recently completed a Spanish language course and if I could utilise their skills, arrangements would be made to get them south to join my team. However, firstly I had to give the RAF sponsors an assurance that when we eventually disembarked *Intrepid*, I would not include these six specialists as members of the STD as they had not been trained to operate in a combatant role; naturally I had to decline the offer.

Another area requiring Spanish language desperately lacking in qualified and competent personnel was in interrogation of prisoners. When I first transferred to the Intelligence Corps, all recruits were trained in escape and evasion and interrogation (normally as the one being interrogated) but this was phased out of the basic training program and current members of the STD were not trained in these areas. Within the first week of hostilities a crashed Argentinian pilot had been rescued from the ocean. Badly injured, he was brought onto *Intrepid* to receive medical treatment.

Once on board and making a satisfactory recovery, someone in their wisdom thought this PoW should be interrogated, and because the STD had trained Spanish linguists they should provide the interrogator. I couldn't do the job because of my linguistic limitations so I delegated Corporal Jock to do the best he could. Jock went to the sick bay but returned shortly to tell me that the prisoner would not communicate with him. Corporal Jock was an avid supporter of Scottish football and although Scotland had not qualified for the last sixteen in the World Cup, he knew a lot about English and Argentinian football, so I briefed him to return to the PoW, initially engage him in conversation relating to football, and if cooperative, extend his enquiries to included details of a military nature. Corporal Jock returned from his interrogation duties with some very useful information regarding the enemy.

Since the establishment of a bridgehead in San Carlos Bay by the Task Force, the Argentinian radio networks serving their ground forces on the Falkland Islands were active almost 24 hours a day. With continuous monitoring of this network, we were able to provide HQ 3 Brigade with up-to-date, in some cases instantaneous, reports on the enemy's activities. During the early part of hostilities and before the Task Force was able to gain total air superiority, the quickest way for Argentinian forces to travel from their headquarters in Stanley to subordinate headquarters located throughout the Islands was by air, usually helicopters.

Fortunately for the STD, most days the Argentinian headquarters at Stanley would notify the outstations of schedules for helicopter flights to and from forward locations. This schedule, in addition to forward locations to be visited, would also include ETA and departure and the route to be taken by the aircraft from the Race Track in Stanley. Immediately the signals containing these aircraft schedules and itineraries were intercepted, the information was passed to various addressees, including the EW cell on *Hermes* for counter-action to be taken. The STD was aware that on some occasions this early warning had enabled our own aircraft to be tasked against these airborne targets and in some cases rendered them well and truly un-airworthy.

When the Argentinians first landed on the Falkland Islands in early April, they had the use of a vessel, the ARA *Bahia Buen Suceso*, a 5,000-ton fleet transport vessel serving as a logistic ship, intended to re-supply the scattered

Argentinian garrison around the Islands. The ship continued to be used by the Argentinians after the commencement of hostilities, that was until the STD intercepted material indicating the ship was to make a voyage from Stanley to Port Howard with estimated times of arrival at various locations along the route. As with the warnings of aircraft movements this information was passed immediately to HMS *Hermes*. The STD heard later that as a result of their tip off, the ship, when moored at Fox Bay, had been located by the British Forces, fired on and suffered extensive damage. Later we found out that the ship had broken from its moorings at Fox Bay and had run aground.

Life for a soldier on board a navy ship was much more comfortable than roughing it on land. It was warm, dry, all meals were prepared by other people and there was a comfortable bed to sleep in. The only thing not available was a bath or shower, but there was hot and cold running water to wash dirty bodies as well as dirty clothes. Wearing the same clothes continuously for five days from 20 May, I knew, without my best friend telling me, that I was in urgent need of a wash and clean clothes so I set out to find a bucket or large bowl so that if I sat in a bath I could rinse myself all over. Actually I went one better and found a shower where the taps had not been removed; believing my luck would run out when I turned the tap on, to my surprise out came warm water so back to my cabin for soap and towel. Under the shower I went and in quick time had managed to cover myself in a thick soapy lather, then, something I had not bargained for, above the sound of the gushing shower water came the early warning siren announcing an incoming air raid. My immediate thought was to return to the EW cabin, so without rinsing off the soap or bothering to dry myself, I hurriedly dressed and returned to my duty post. Appearing a little wet and bedraggled I had some explaining to do to those members on duty.

A couple of days later I decided to try my luck again. Like the first time, I got into the shower, lathered all over and again came the early warning siren. This time I reacted differently. I carried on showering, although rather more hurriedly than usual, and dried myself thoroughly in the belief that if the ship was hit and I was killed, it didn't really matter if I was dressed or not. On leaving the shower cubicle I furtively checked the outside passages to ensure I would not be observed, and in doing so I noticed the Roman Catholic padre walking away from me. Thinking it would be all right I followed him, to hear him enquire 'Is that you David?' and when I declared myself he said he knew I had been in the shower and had been praying for me! Apparently the Padre's cabin was adjacent to the shower cubicle, and it would appear that he and several others had observed me entering the shower room both times, and had politely remained in the outside passage until I had completed my ablutions.

In addition to the Catholic padre there was also a Protestant padre by the name of Michael on board *Intrepid*. On our journey south, Michael and members of the STD were frequently seen engaged in friendly conversations and he became popular with all members – it was he who introduced the word 'biff' into our vocabulary by telling us stories of how Marines were frequently discovered alone in the area set aside as the prayer room, thinking they would not be discovered. The solitary practice ceased when Michael declared the prayer room only open at specified times during the day, at all other times it was kept locked.

Religion has never been an important issue in my life since I was a young apprentice soldier, when almost every Sunday morning over a three-year period I was forced into my best uniform that had taken me several hours the previous day to clean and polish to inspection standard, ordered onto a parade ground, inspected, then marched to Church behind the regimental band. On arrival at the church door, the Catholics who had attended their own church service earlier in the morning and non-Christians, in the main the few Burmese apprentices who were mostly Theravada Buddhists, were dismissed and allowed to return to their barrack rooms. Those agnostic, atheists and 'can't be bothered with religion' apprentices not wishing to participate in the Church service, were seated at the back. They had to remain assembled until the end of the service when all the apprentices reformed in smaller units and were marched back to their barracks.

This system of enforcing young men to attend Church was, in addition to being sanctioned by Military authorities, also condoned by the Church of England. As a teenager I could think of better things to be doing with my time for the whole of a Sunday morning. However, since those formative years, I have frequently debated the whys and wherefores of religious issues with friends and have often asked if, as a regular soldier, in the event of war and finding myself in the position of looking down the barrel of an enemy rifle, I would pray.

Fortunately for me, throughout the entire period of hostilities, this question was never answered. What did happen was that one morning while in Bomb Alley I decided to go to the ward room on the scrounge for a cup of coffee. When I got there I found that Padre Michael was doing the same thing. Not long after we poured our coffee the air raid early warning siren was activated. Thinking I would allow those who manned the guns to get to their duty stations before I returned to the EW cell, I sat in an armchair to finish my drink. In contrast, Michael was running in small circles in the centre of the ward room with a petrified expression on his face. Instructions were issued to 'hit the deck', and as I lay on the floor I observed that Michael had failed to carry out the order. I shouted to him to get down which he did immediately and I crawled to his side. Trying to reassure and comfort him I managed to get him from the centre of the ward room and into a chair at the side where I remained with him until the all clear. After a cup of tea and time to regain his composure, Michael was once again his assured and calm self and asked me what had happened after he heard the air raid warning siren. I explained to him and asked him where he normally went during air raids. He replied that during the day he spent a great deal of time in the sick bay and when the alarm was activated he would go into a small two-berth cabin and lie on one of the bunks until the all clear. It then became apparent that he had panicked because for the first time he was caught out in the open and away from the security of his small cabin environment. So in this instance I gave reassurance and comfort to a man of the cloth and not the other way around.

On 27 May *Intrepid* – or more specifically AS Neil Wilkinson, aimer of the starboard Bofors gun – struck a blow, though he could not be sure if it had landed until arrival back in England. Late in the afternoon two Skyhawks appeared. According to an interview with him (and with the pilot) in *Classic Aircraft* magazine (May 2010) 'I managed to fire six rounds off at the two Skyhawks and saw one of them go over the hill with smoke trailing from it.' The pilot was Mariano Velasco, the man who caused the most damage to British forces in a single attack during the conflict by hitting and sinking HMS *Coventry* with three 500lb bombs just two days before. Comodoro Velasco ejected safely over West Falkand (Gran Malvina) and was eventually transferred to the hospital ship *Bahia Paraiso*.

Fighting a war from a seaborne platform means re-supply – of fuel, water, rations, weapons and other expendable stores and equipment. The Replenishment at Sea (RAS) for the Royal Navy is the responsibility of the Royal Fleet Auxiliary (RFA), a civilian-manned flotilla, highly trained in the very hazardous jobs of transferring fuel in rubber hoses by the abeam method and heavy jackstay method to transfer stores. Having spent a couple of weeks in her operational state in the waters of the Falkland Islands it became necessary for HMS *Intrepid* to be replenished. Because it is was far from the safest place to carry out a RAS at anchor within the waters of San Carlos Bay, or even, for that matter, within the 'total exclusion zone' created around the Falkland Islands, for fear of representing an easy target to incoming hostile enemy aircraft, it was necessary for *Intrepid* to temporarily depart the war zone for safer waters and out of range of incoming enemy aircraft. Having sailed through the night the crew and embarked force of *Intrepid* awoke on a Sunday morning to find themselves in relatively calm waters and in the belief that, at least for a short while, the ship was out of range of enemy aircraft. The RFA ships rendezvoused on time and with the customary speed and professionalism expected of them soon had supply lines between them and *Intrepid* to begin the task of re-supply of fuel and stores. Meanwhile, apart from those members on essential duty, the crew took advantage of this break in the war to catch up on personal chores and an air of relaxation descended over the ship.

After lunch a film show for all ranks had been arranged in the ward room. Prior to the film starting every tiny space within the ward room had been occupied. I managed to get a seat on top of a table along the starboard side of the ship and farthest away from the screen. About halfway through during a lull in the sound track and at about the time when half the audience were already asleep, with the remainder thinking of doing so, the relative peace was broken by a loud sound similar to 'whoosh, whoosh'. This was an unfamiliar sound to me, but a member of the audience suddenly shouted 'chaff, chaff!' Bearing in mind the makeshift cinema was in relative darkness, such was the sudden exodus that by the time someone had found the light switch, the room was empty apart from about half a dozen members of the audience sat at the back on the tables.

Chaff (radar countermeasures made from aluminium) in appearance looks like finely shredded tin foil and when fired from its container explodes to spread upwards and outwards to form a very large cloud effect. The use of chaff is primarily against detection by missiles; if the missile operator has not 'locked on' the missile, it will give him a picture of a cluster of secondary targets or swamp his screen. On this occasion when the chaff was fired from HMS *Intrepid's* nine-barrelled launchers, one of the duty personnel on whom the safety of the ship relied, in the belief that the ship was about to be attacked from the air, fired the chaff before the call to report to duty stations could be given. Having fired the chaff, it was an extremely short period of time before HMS *Intrepid* had carried out an emergency break away from its RAS and was attempting to put space between it and the RFA re-supply ships.

The cinema audience never did see the end of the film and the talk for the rest of the day was about what possible damage had been done to the ship in carrying out an emergency break away. Many of the ship's crew believed this to be the first such instance since the end of the Second World War, and talked of the dreaded Exocet missiles and what might have been. Personally I believed the incident to have been one of mistaken identity. The location for the RAS was too far away from the Argentine Air Force safe operational area and when returning to the EW cell, analyses of activity intercepted that afternoon by the STD gave no indication of aerial activity in close proximity to the RAS. The incident however, was responsible for a rapid change from a relaxed cruising environment to one of a professional military unit prepared once again to enter the war zone.

10

THE BATTLE FOR GOOSE GREEN

History has recorded that on the establishment of a bridgehead by 3 Brigade in San Carlos Bay, the break out was by the Marines and 3 Para 'yomping' in a northerly direction towards Stanley and 2 Para moving in a southerly direction towards Stanley via Goose Green. Early interception of the Argentinian Ground Forces' radio communication suggested the Marine and 3 Para advance would be virtually unopposed, while 2 Para would possibly be confronted by the enemy's 'Mercedes Task Force' in the Goose Green/Darwin area.

On 26 May 1982 in response to a knock on the EW cell door there stood OC B Coy and another officer, who the former introduced as Lieutenant Colonel 'H' Jones, CO 2 Para and asked if I would brief the Colonel on the latest enemy situation because I was one of the few involved in intelligence gathering that had knowledge of the current situation concerning Argentinian ground forces. My immediate response was to ask if Colonel Jones had the personal security clearances to allow him access to my product. This did not go down well with OC B Coy but I explained to him that prior to leaving the UK, I was told in no uncertain manner that 'peacetime restraints' were the order of the day and no clearance meant no briefing. OC B Coy's reaction was to grab me by the neck, push me against the cabin wall and inform me we were not playing at being soldiers but this in fact was a real war. I was entirely in agreement with him and, in my defence, was only following orders. When OC B Coy explained to me that Colonel 'H' was to lead his battalion into a battle for Goose Green, I ignored instructions about 'peacetime restraints' and asked Colonel 'H' if he would like to come in for a briefing.

As part of the 2 Para advance on Goose Green pre-briefing process, Colonel 'H' had attended an intelligence briefing in the intelligence cell of 3 Brigade on board HMS *Fearless* before arriving on HMS *Intrepid*. Prior to the start of my briefing, I noticed Colonel 'H' scrutinising a large map on the wall of the cell on which were symbols, markings, comments and notes of points of interest as reported by the STD to our sponsors back in the UK and to the command headquarters we had been supporting since the commencement of hostilities. In particular he was looking at locations gleaned from the Argentinian command and control network earlier in the morning. On taking out his notebook, in which I saw were written the same grid references as those shown on the map, he asked if the locations on the map represented those he had seen on the map board in 3 Brigade intelligence cell during his briefing visit earlier in the morning. I confirmed they were and that they represented only part of a monitored conversation because broadcasting conditions had been very poor at the time and a great deal of the conversation had been missed.

I explained to Colonel 'H' that at the time of intercept, apart from there being ten or twelve grid references gleaned from the enemy's fragmented conversation between themselves, in isolation they did not mean much to me. However, in the much broader intelligence picture and because they were used in the context of location points, they might be of value and fit part of a jigsaw. Therefore I forwarded the details to the Brigade intelligence cell. The response was to inform me that the grid references I had forwarded earlier were of British locations and the intelligence cell was well aware of where their own forces were located. I was dumbstruck; here I was passing them real-time information taken from an enemy's communications that indicated the enemy possibly knew the locations of the British Ground Forces – and apparently they didn't realise the significance of what I was giving them.

During our conversation Colonel 'H' had not taken his eyes from the map then when I had finished talking his response was 'Oh bloody hell' and was it possible for me to confirm the accuracy of the references. I showed him the raw intercept on which the report was based and had a linguist carry out a second translation of the content. Colonel 'H' then carefully rechecked the grid references from the second translation with those on his piece of paper. Finding no discrepancies he then rechecked his figures with those on the wall map and then replied, 'Do you know what I think, these grid references are enemy locations, not British locations, I can confirm that elements of my battalion and other British forces are based in the vicinity of grid reference one, two, five and seven but are definitely not anywhere near the remaining grids.' The grid references I had forwarded to the intelligence cell were Argentinian locations; it is my belief that what most likely happened was the Brigade intelligence cell 'duty plotter' started to put the locations on the map and because the first few were known locations of British forces, wrongly assumed that the other locations were also.

Continuing with my briefing we then discussed estimated strengths of the Argentinian ground forces located in and around the Goose Green/ Darwin area. From the latest strength returns forwarded by units in the area under question back to Argentine headquarters in Stanley, I informed Colonel 'H' that for the past seven to ten days the combined strength had always been in the region of 1,500 men. The vast majority were army personnel, comprising a large element of 12 Infantry Regiment, supported by smaller elements from an airborne infantry regiment and Special Forces group. In addition to its ground forces, the Argentinians also had a rough estimate of some 200 Air Force specialists in support of the co-located Pucara aircraft base.

The briefing had been progressing well with Colonel 'H's' active and enquiring mind frequently requesting the answers to pertinent questions, but then came the question those in the field of gathering intelligence dread: 'Can you assess the situation and give an indication as to my success or otherwise if I was to mount an attack on the enemy positions?' My response was that quite simply the role of the STD during Operation *Corporate* was to gather Sigint on the enemy and forward this to initially 3 Brigade, later to 5 Brigade and HQ Land Forces Falkland Islands, to enable their intelligence cells to include this data, or otherwise, in the overall assessment of the Argentinian Forces located on or in the vicinity of the Falkland Islands. In this particular operational role if the STD were to try to assess the situation the end product would suffer from the effects of tunnel vision because its assessment would be based on one area of intelligence and would be further distorted by the fact such assessments based on this field of intelligence had little or no content regarding the activities or locations of our own forces.

This notwithstanding, Colonel 'H' was a very persuasive officer and before long we were discussing tactics along with the enemy's known strengths and weaknesses. Not attending Staff College I was ignorant of the finer points as to numbers of troops and types of equipment an attacking force should have over a defending force, and not any old defending force but one that had been 'dug in' for a considerable period of time. I was aware the British Army doctrine was for any chance of success an attacking force must be superior in strength by a ratio of three or four to one. I asked Colonel 'H' what his overall strength at the time of engagement was likely to be; he replied that he had almost a full battalion of Paras plus supporting arms and services, that in all would amount to a little over 900 men; far too few for a realistic and successful outcome. Our conversation then turned to calibre, skill and professionalism of the opposing forces and their men under command; here I believed, there was no real comparison. To distinguish between the opposing forces I likened it to football; the British soldier was top of the Premier Division while the Argentinian soldier played in the local Sunday pub league.

According to data collected by the STD it was apparent that a great many Argentinian soldiers were young conscripts enlisted from senior schools, given the minimum of training, then by night deployed to a combat area they believed to be in the Argentine, where they were ordered to fight in the defence of their country. In reality and without knowing it, they had been flown in excess of 400 miles to the Falkland Islands to fight against the British Army. With such a vast difference in capability between the ground forces, it could quite easily be perceived that a superiorly trained minority could on this occasion overcome a force greater in numbers but lacking in training and experience. On conclusion of the briefing Colonel 'H' said in view of what had been said he now felt that his battalion's chances of success were greater than at first expected and he would go and prepare for engagement with the enemy.

The day following the Colonel's briefing it was obvious from the messages flowing fast and furious between the outstation at Goose Green and control in Stanley that a battle between two opposing ground forces had begun. On the morning of 28 May, the VHF frequency spectrum on the automated scanner became active. This was the very first intercept by the STD of Argentinian VHF voice communications since the landing of the Task Force. When it was established that the intercepted communications were emanating from the Goose Green/Darwin area, we in the EW cell were surprised to say the least because we had considered the distance from the battle area to HMS *Intrepid* at anchorage in San Carlos Bay to be too far to intercept signals from the small battlefield radios in use by the forward Argentinian troops. We were not complaining because it enabled us to maintain continuity on the enemy's actions and to forward instant state of play and enemy situation reports to the intelligence cell at 3 Brigade.

By the afternoon and having all available members of the STD report for duty in the EW cell to ensure maximum manning of all intercept equipments, through a thorough and painstaking analysis of the intercepted Argentinian communications, we were aware that while the ongoing battle for Goose Green/Darwin between Argentinian ground forces and British ground forces comprising 2 Para and supporting elements was in progress, Argentinian reinforcements had been flown in from Stanley by helicopter. It was obvious that the Argentinian helicopter pilots had no intentions of getting involved in the ground battle as they landed their helicopters some distance from where the fighting was actually taking place.

On landing and disembarking their passengers, the helicopters immediately left to return to Stanley to bring additional reinforcements forward to provide extra fire power and support to those already engaged in battle. Before the second wave of Argentinian reinforcements had arrived, those reinforcements in the first wave were running around like headless chickens. Having organised themselves to give assistance to their fellow countrymen in the ongoing battle, to their commanders' horror and dismay, they discovered that they had left their base without maps or compasses and had no way of orientating themselves on the ground. One field commander contacted his headquarters in Stanley by radio for directions to the battle. The commander at headquarters replied that surely the forward commander could hear the sound of the battle and to move forward in the direction of where the sound was coming from, to which the commander of the reinforcements replied he was unable to accurately pin point the position of his own ground forces because of small arms gunfire, mortars and shells fired by the enemy artillery and navy out at sea – the noise of battle was coming from 360 degrees around him.

Eventually the Argentinian ground forces headquarters in Stanley, via the radio communications link being monitored by the STD, with the aid of their own map and awareness of the current situation tried to determine the location on the ground of the forward field commander in order to direct him and his supporting elements in the right direction. Reports relating to the Argentinian reinforcements having been brought forward from rear echelons were constantly updated by the STD and forwarded to the intelligence cell on HMS *Fearless*.

I mention this incident in particular because, having monitored the Argentinian communications between Stanley and Goose Green concerning action to be taken by the Argentine forces at the time of the ceasefire and eventual surrender, the STD was aware that at the time of the ceasefire, all Argentine forces in the vicinity of Goose Green had been directed to the area of a large sheep shed where they were to remain until the surrender documents had been signed. At first light the Argentinians were assembled for a head count resulting in the estimated figure of 1,500, as given to Colonel 'H', which including the Argentinian dead, increased by the reinforcements to approximately 2,500. But our intercept suggested that none of the possible 1,000 reinforcements was ever committed to battle. Once all reinforcements had reached the combat zone they remained at their point of arrival and waited until cover of darkness before moving forward to join their own forces.

Shortly after the signing of the surrender documents, through another British source, the total figures of those Argentinians taken prisoner or presumed dead were received at 3 Brigade. Not long after I had the intelligence cell 3 Brigade on the radio asking why we had been so inaccurate with the previous manning figures for the Goose Green/Darwin area. I informed them that I stood by the figures as previously passed and suggested they look at our intelligence reports and summaries from the day before, informing 3 Brigade and others that reinforcements had arrived and this was the reason for the increase in numbers at Goose Green. The Brigade intelligence cell operator replied 'What reports about reinforcements?'

Remember, the STD had at this time, in addition to the facility of routing encrypted reports to our customers electronically via the Royal Navy lines of secure communications, a direct one-to-one secure voice radio link to the intelligence cell 3 Brigade, with all messages passed over this link logged and recorded. I know that critical intelligence relating to the arrival of Argentinian reinforcements during the battle of Goose Green/Darwin was passed to the intelligence cell, in real time. What happened on its receipt I will never know. At a push I could believe it possible that reports originated by the STD and transmitted via the Royal Navy secure communication systems may not have been received by all addressees. Through experience I was aware that, without any breaches of security, this happened on the very rarest of occasions. I will never accept that the intelligence cell of 3 Brigade were not verbally informed of the Argentinian ground forces reinforcements to Goose Green in real time and as events unfolded because the majority of these intelligence situation reports were personally transmitted by myself.

Colonel 'H', a very courageous man, was killed on Darwin Hill in the battle for Goose Green and he was posthumously awarded the Victoria Cross. The 2I/C of 2 Para, Major Chris Keeble, took command of 2 Para until the arrival on 2 June, by parachute, of Lieutenant Colonel David Chandler. I know nothing of Major Chris Keeble but find it very strange why he was never given Command of 2 Para after he led them to victory over the Argentinians at Goose Green.

Several days after the battle for Goose Green, I was once again sat in the ward room of HMS *Intrepid*, this time in the company of two young officers from 2 Para. On noting that my parent unit was the Intelligence Corps, it wasn't long before the conversation turned to intelligence or, in their case, the lack of it. One of the Para officers said that by appointment he was a Company Intelligence Officer, but pointed to the notepad on his knee which had nothing on it and informed me that was the total sum of intelligence he received throughout the battle for Goose Green.

All I could say was that it was a good job the STD's intelligence reports had been transmitted in real time, as and when events happened via a variety of reporting vehicles and while I had to accept intelligence wasn't passed direct to the troops on the ground by the STD, at least the content of our intelligence reports would have been included in other reports, apart from those sent direct by the STD to the Brigade intelligence cell on HMS *Fearless*. It was assumed the Task Force would have been up to date on events as they happened.

On the cessation of hostilities, amongst a limited assortment of Argentinian battlefield communication equipments that had been confiscated and stored in the belief that the items appeared to be in working order, were a number of very small handheld voice communicating (similar to a walkie-talkie) units, capable of operating in the VHF band with a transmitting distance of about 8–10 miles. The handheld radios we discovered were similar to those used by the Argentinian units during the battle for Goose Green and may have been the actual ones. How fortunate we were to intercept the transmissions from these radios some 30–40 miles away, allowing the STD to hear and follow the battle as fought by the Argentinians. These unusual circumstances go to show that the performance of radio communications should never be underestimated, nor for that matter taken for granted; even under perfect conditions a radio operator would not conceive it possible for radio signals to travel in excess of 40 miles when transmitted from a radio designed to have a range of 10 miles maximum.

Since the British success at Goose Green, I have frequently wondered what might have happened if I had carried out instructions not to pass on our product to those not having the mandatory security vetting clearances to receive such information. What if Colonel 'H' had not been so inquisitive about the locations on the map within the STD, and what if I had stuck to the fact that we were deployed as an intelligence collecting platform and not as a means whereby a current assessment could be made without the referral and possible inclusion of collateral evidence and other forms of intelligence? Would the outcome have been different?

11

ARRIVAL OF 5 BRIGADE

Prior to the arrival of the Task Force in San Carlos Bay, it was deemed necessary to increase the strength of the Land Task Force if there was to be any chance of a successful end to hostilities, but from where were these reinforcements to be taken? Although not a single shot had been fired in anger between the NATO forces located in West Germany and the Cold War enemy in East Germany and beyond since 1945, it would have been less of a problem to extract blood from a stone than temporarily rob Britain's frontline defences in the British Army of the Rhine. The Army's 'Ace Mobile Force' stationed near Salisbury was also not available because it was a designated NATO force.

With very few options remaining to provide reinforcements, it was decided to send 5 Brigade, consisting of the last remaining original battalion, the 1st/7th (Duke of Edinburgh's Own) Gurkha Rifles, because the Brigade's other two battalions – 2 and 3 Para – were already on active service with 3 Brigade. It was reinforced with the 1st Welsh Guards whose previous role was mainly to provide ceremonial duties at Windsor Castle and terrorism and hijack duties at Heathrow Airport, and 2nd Scots Guards, who had for some considerable time been providing guardsmen for ceremonial duties at Buckingham Palace.

With the size of the Land Force now increased to eight infantry battalions – the largest British force sent into conventional military action since the Second World War – the appointment of a new Land Force Commander was essential. It was thought that the Army would provide an experienced divisional commander and staff, but this was not to be. On 31 May Major General Jeremy Moore, previously Commander Commando Forces and military adviser to Admiral Fieldhouse, was the designated Commander Land Forces Falkland Islands (LFFI); thus the Task Force continued to be under the control of the Royal Navy.

Transporting 5 Brigade and the headquarters staff of the Land Forces to the war zone was undertaken by the luxurious passenger liner *Queen Elizabeth 2nd* (QE2), and the Stena roll-on-roll-off ships *Baltic Ferry* and *Nordic Ferry*. The ferries departed on 9 May, followed by the *QE2* on 12 May 1982. On arrival at Grytviken, South Georgia on 27 May, to avoid any damage or possible loss to the *QE2* and the other two vessels, the Gurkha Battalion was cross decked to the *Norland* and the Scots and Welsh Guards with brigade-supporting arms were cross decked to the *Canberra*. The *QE2* and both ferries were turned around, returning to the UK and a hero's welcome.

Landing at San Carlos Bay on 31 May, Major General Moore established a headquarters on HMS *Fearless* and assumed the overall responsibility for LFFI. Increasing the LFFI from one to two brigades should have included the provision by the STD of Sigint support to 5 Brigade. This proved to be impossible because with limited stocks of radio equipment our secure radio voice link to the intelligence cell of 3 Brigade had to be withdrawn and reissued to others with a so-called higher priority than ourselves.

With the loss of our own dedicated radio equipment it meant the end of the STD's real time reporting role of enemy flight and troop movements within and around the Falkland Islands. However, all was not lost, with the increase in size of the intelligence cell in the operations room of HMS *Fearless*, one member of staff within the HQ LFFI who travelled south on the *QE2* was Major Mike Dawkins, Intelligence Corps, whose appointment was that of EW Liaison Officer. This was a similar role to one I had played many years previously when serving with 14 Signal Regiment EW and deployed with 1st British Corps on FTXs. I first met Mike in Cyprus in 1959, and having frequently served together, we knew each other very well. While it was impossible to work directly with the EW Liaison Officer on *Fearless*, it was discovered that the quickest means possible for the STD's operational product to reach him was by a formal secure message sent via Arrowflex over a satellite link to one of our sponsors in England and for them to reroute by other means direct to *Fearless*. This way, most communications could be on the EW Liaison Officer's desk within eight minutes, the signals having travelled about 18,000 miles. Not quite the speed the STD had been accustomed to but the next best thing to almost real time reporting.

By the time 5 Brigade, under the command of Brigadier Tony Wilson, had arrived in San Carlos Bay, operational plans for the continuation of the advance onto Stanley had been confirmed. With the break out from the bridgehead by the majority of units of 3 Brigade to advance to Stanley in a northerly direction and 2 Para advancing south via Goose Green, it was planned for 5 Brigade to follow a similar route as 2 Para. On receipt of their orders to advance, 5 Brigade automatically assumed they would be moved to their forward locations by some form of transportation. In fact the Scots and Welsh Guards were informed that transportation was virtually impossible as all available cargo and troop-carrying vessels had been committed to loading stores and equipment from the base at San Carlos to a forward location; they, like others before them, would have to walk carrying all their personal kit.

There was some considerable objection to this order because having spent large amounts of time on ceremonial duties prior to their departure from the UK, the Guards were not physically fit enough as they had had only limited time to train and prepare themselves. On the other hand, the Gurkhas, who had been committed to a different peace time role and were able to maintain a reasonable standard of military fitness, didn't need asking twice. They secured their kit and weapons on their backs and immediately set off to 'yomp' in the direction of Goose Green/Darwin to relieve 2 Para who were waiting to advance towards Fitzroy. Eventually, after much time spent on discussing how the Guards battalions were to deploy forward, an agreement was reached whereby all the recently loaded stores and equipment on board HMS *Fearless* and *Intrepid* would be unloaded and replaced with a human cargo consisting of the Welsh Guards on *Fearless* and the Scots Guards on *Intrepid*.

During the night of 6/7 June, HMS *Fearless* with an embarked force from 1st Welsh Guards sailed in extreme weather conditions south through the Falkland Sound, continuing around south of East Island and into Choiseul Sound to a location in the vicinity of Fitzroy. Unfortunately, on arrival at the designated location the Battalion was confronted by a shortage of LCUs to move the troops from *Fearless* onto land and only about half of the Battalion were able to disembark. The remainder were taken to Goose Green where the following morning they were picked up by the RFA *Sir*

Galahad to be taken to Bluff Cove. On the way the *Sir Galahad* made a transit stop at Fitzroy to unload elements of the Rapier missile and Field Ambulance units.

At the time it was rumoured that the elements of the Welsh Guards on board the *Sir Galahad* were invited to embark and make the remaining journey to Bluff Cove on foot, a distance of some five miles as the crow flies from Fitzroy. This offer was declined because apparently the Argentinians had destroyed a bridge over a narrow inlet between Fitzroy and Bluff Cove, which would have meant an approximately 20-mile detour to reach their new forward location. Instead they chose to remain in the most vulnerable of positions, on an auxiliary supply ship, which apart from her 2 40mm Bofors AA guns was not equipped with weaponry to defend itself or to retaliate in a hostile situation without the added protection of a Royal Navy warship. She was at anchor offshore and in full view of approaching enemy aircraft. The Argentinians exploited *Sir Galahad's* vulnerability; it was sunk on the morning of 8 June 1982, with 50 members of 5 Brigade either killed or missing. Over the years I have wondered how things might have gone differently; what if:

a. Rapier missile system had been deployed at time of arrival the night before to give anti-aircraft protection. Basic standard operating procedures I would have thought.

b. Troops had been put ashore under cover of darkness the night before instead of waiting during five hours of daylight before ordered to do so. Unlike the initial 3 Brigade landing where the Brigade was entering into the relative unknown territory, the Guards had the advantage of knowing that friendly forces already occupied the immediate area.

c. Priority had been given to get men ashore before stores and equipment.

d. The Welsh Guards had been instructed to adhere to the disciplines of dress and behaviour in a war environment i.e. kitted and booted at all times and not watching videos, playing cards etc. Some time after hostilities had ceased and I had returned to my home duty station, I watched a television programme depicting the sinking of the *Sir Galahad* in which scenes clearly showed the unprepared state of the embarked force while in a battle zone.

e. The rules relating to the wearing of anti flash clothing had been extended to members of an embarked force and not only the ship's crew.

f. A location for a bridgehead out of view of the enemy had been chosen; Bluff Cove would have been in full view of enemy troops dug in on Mount Harriet, which was only some 15km away.

A short time after the loss of the *Sir Galahad*, two members of her crew fortunate enough to have survived the attack were given temporary accommodation in the same four-man cabin as myself on board *Intrepid*. They arrived with only the clothes they were wearing at the time of the aerial attack, and the smell of diesel oil and smoke left no one in any doubt as to where they had been. The two crew members remained in the cabin for about two days while they re-equipped and rested after their ordeal. During our infrequent social contact during this time, I gleaned they were not impressed with the way the situation was allowed to develop and the events that followed immediately after their ship was hit by enemy bombs.

The *Sir Galahad* had been offered as a target of opportunity to the Argentinian Air Force. From intercepted air activity it was most probable that the aircraft were en route to complete a different mission but because the activities of the British within the area of Bluff Cove had been observed by Argentinian ground forces located a short distance away, the original tasking was aborted in favour of the highly vulnerable target. It is also believed that from enemy intercepted material available at the time the reason to abort the original target mission was because it could not be located. Fearful of spending more time attempting to locate the designated target, with the resulting problem of running low on fuel and not making it back to base, they left the projected target mission area and it was only while returning to their base that they received the orders to attack *Sir Galahad*.

Three days after their arrival in San Carlos, time spent adjusting to life in a hostile environment after the luxury of sailing on a cruise liner, on the night of 5/6 June the Scots Guards embarked on *Intrepid* for onward transportation from San Carlos to the entrance of Choiseul Sound. There they cross decked to waiting LCUs to land at Lively Island for the first leg of their journey to relieve 2 Para at Bluff Cove.

It became evident, in my opinion, that elements of 5 Brigade were not of the same calibre as 3 Brigade; the behaviour of a few revealed that the brigade had not been subjected to the same discipline and management during their journey south for entry into a war zone. Throughout the entire journey from England to the Falkland Islands I was not aware of any disciplinary problems from the embarked force sailing on *Intrepid*. In fact I felt quite proud to be counted amongst them because I considered their overall behaviour and conduct a fine example of a professional armed force. But in the course of one night my pride in association was lessened owing to the antics of a minority hooligan element of the Scots Guards ripping notice boards from walls, activating fire extinguishers and drinking too much alcohol.

By their behaviour some of the junior officers had been as bad as the rank and file. Because I had become well known to the crew of *Intrepid* through being on their ship for such a long time, I was approached by a crew member to assist him in removing three or four young officers from the ward room. These officers had finished their evening meal, then unlike the other officers on board who by this time had returned to their operational duties, insisted on delaying the ward room staff by ordering rounds of sherry and port. When I arrived the subalterns were in a state of undress, having removed their outer garments and boots, and by the noise they were making I guessed they had consumed a fair amount of alcohol. However, a few words from me – at this time I had not taken off my boots and had lived in the same combat clothing day and night for the past three weeks – soon sent them back to look after their troops, allowing the ward room stewards to return to action stations.

The day after the Scots Guards left HMS *Intrepid* was the day we in the STD discovered all our winter warfare clothing had been stolen from the back of our locked Land Rovers. After my years of service with 24/7 operational units and training exercises in Germany, this very short experience of working with troops taken from ceremonial duties to be employed on active service with other members of the armed forces is not one that I would recommend. Apparently some members of the embarked civilian press corps, normally in a hurry to file stories of this nature, decided amongst themselves that the night's antisocial behaviour should go unreported. Because the culprits were drawn from a very small minority it would not be good for the morale of the majority to have this type of story plastered over the front pages of English newspapers. Thank you to the embarked Press Corps.

I have already mentioned that with the arrival of 5 Brigade and the removal of the dedicated secure VHF radio equipment for higher priority users, the STD on HMS *Intrepid* and the EW Liaison Officer Major Mike

Dawkins on HMS *Fearless* did not have any direct means of communication, but there were several occasions when my presence on board *Fearless* was required to attended briefings. On such occasions it would be when *Fearless* and *Intrepid* were both at anchor in San Carlos Bay. My transport to attend such meetings would be whatever was available at the time, either by sea (rigid raider) or air (helicopter). All transportation was on an ad hoc basis. I often had to stand on the flight deck of *Intrepid* for several hours before the start of the meeting I was to attend and hitch a ride, sometimes resorting to pleading, on the first transport heading my way whose eventual destination would be *Fearless*; sometimes this was via several stages on land or sea platforms. Usually I would be in possession of highly classified codeword reports which, because we did not have a recognised secure courier bag or briefcase, would be double wrapped inside two standard paper envelopes.

On one such visit to HMS *Fearless* to attend an intelligence meeting, the only means of transport available was a helicopter, but as the pilot was running late on his multi stopover round trip, the load master informed me that the pilot couldn't land on the flight deck and so long as I didn't mind, I would have to be dropped by winch. This didn't worry me because I had been trained in alighting by rope from helicopters and I had already descended by winch on a couple of occasions while at anchorage in San Carlos Bay.

Once the helicopter was airborne, I was fitted into a harness and 'once only suit'. With *Fearless* below the helicopter, the winch man began lowering me onto the flight deck below. I had not descended far before I was aware that ships at anchor were sounding their horns as an early warning against an approaching attack by enemy air craft – at this stage of the war practice alarm warnings were a thing of the past, every time the ship's horn was sounded, the situation was for real. The winch man shouted that he didn't have time to reel me in as the pilot was heading towards the high ground around the bay, in order to provide him limited aerial protection against the incoming enemy aircraft. While all this flight activity was in progress I was left hanging clutching my envelope of highly classified material. At the same time, from several directions, the enemy air force dropped their bombs and fired on the ships at anchor in the bay. Hanging from the winch line I had a clear view into the cockpit of the enemy aircraft as they flew close by, but the only thing I could think about was what if I was to lose my envelope? Eventually the helicopter returned to *Fearless* where I was lowered by winch – you would have thought that after this ordeal they would have had the decency to land on the flight deck – and I was able to attend my meeting with my classified material still intact and perfectly dry. My return to *Intrepid* was by rigid raider and thankfully not another helicopter.

Shortly after 5 Brigade's breakout from the beachhead at San Carlos Bay, I received a request from the Y Tp via the intelligence cell on *Fearless* to make available one of the STD's JNCO Spanish linguists to provide steerage on Argentinian radio transmissions in order that a civilian engineer, who had accompanied the Marines to war, could carry out trials on his newly developed radio jamming equipment. At the time of this request we were buried under a backlog of intercepted messages waiting decryption and translation due to a lack of manpower, in particular Spanish linguists. I was being asked to lose one of my valuable assets from a very important operational role in order to help some engineer play with his toys – not likely.

On receipt of the request I immediately contacted my MoD sponsor branch for their assistance in resolving this matter in my favour, but as usual their response was 'you're the person on the spot, do what you think best and we will support your decision.' The engineer's intention was to attempt to jam the Argentinian Command and Control Radio Network, the network that throughout the period of hostilities had been designated the STD's number one priority, as it had provided the Task Force with a considerable amount of very useful intelligence. This was like amputating the hand that feeds you and yet another example of those in command having little or no knowledge of supporting assets. My immediate reaction was to employ delaying tactics; whenever Y Tp asked for the Spanish linguist I would reply that he had recently completed a long period of shift working and was attempting to catch up on lost sleep. This worked and eventually the engineer had to proceed forward with the Task Force for his own protection. I learnt later that the engineer was on *Sir Galahad* when the ship was attacked by enemy aircraft while at anchor in Bluff Cove. If I had allowed the use of one of my linguists, he too would have been on board at the time of the attack. I was informed that while the engineer escaped injury, nothing was known about the fate of his radio jamming equipment.

The request for the STD to provide specialist assistance to allow a civilian engineer to carry out trials on a newly developed piece of equipment in a war environment showed a distinct lack of knowledge of EW. During my two years of duty between 1966 and 1968 with a small independent 1st

British Corps EW unit located at Schafholdendorf, I, together with a few others, gained experience in the skills and use of jamming and intrusion of British Army radio transmissions, an area of expertise that takes some considerable time in which to become a proficient operator. In fact, most of our instruction within this specialist field came via a Canadian Army Signal Regiment SNCO on exchange with the unit. In those days the Canadian Army had a more relaxed attitude towards jamming and intrusion and encouraged its use during most field training exercises.

Before the jamming of a radio station or network can commence, intensive monitoring of the communications must have taken place to ensure a comprehensive knowledge of the equipment to be jammed. Once this has been established the type of jamming to be used must be decided. In their ignorance most people believe it is just a case of reproducing a loud noise through an amplifier via a radio set tuned into the same frequency as the equipment to be jammed.

If this was the case then the jamming site would be vulnerable to direction finding and possible location by the enemy that could result in the jamming equipment being made inoperable. Also it would of course deny any communication intercepting units the ability to maintain continuity of the targeted communications. At the time of my tour of duty with this independent unit the most effective means of jamming transmissions was proven to be through the use of a hair comb. By placing the comb close to a microphone and slowly running a finger nail across the teeth of the comb, a non continuous sound of varying tones and decibel (dB) strengths is made. The sound is reminiscent of natural background noise, with intervals between the sounds sufficient to fragment any voice conversation, thereby making it virtually impossible to pass required data without making several repeats to ensure the listener receives the full message.

The advantage to this type of jamming is that the transmitting operator often becomes frustrated by his inability to pass the required data and will change the content of his transmission in the hope it will be received, often neglecting to adhere to security and other safety precautions relevant to the content of his communication. In addition it also allows any intercepting operator more time to produce a complete copy owing to repetition of the same text. In the light of my own practical experience and knowledge of jamming communication signals one can see why I strongly opposed the use of a member of the STD to accompany a civilian engineer in his quest to prove his invention.

During the early morning of the day before the attack on *Sir Galahad*, HMS *Intrepid* had been employed on duties of a logistic nature, transporting stores and equipment from San Carlos Bay to a forward base area at Fitzroy. On her return to anchorage, WO2 Andy Grey, at the time on duty lending linguistic support to the EW cell on *Intrepid*, realised that the automated intercept receiver had become active. On transferring this new intercepted frequency to his own receiver it was obvious to Andy that he was listening to an Argentinian Air Force voice transmission between two aircraft. That he could listen to their conversation was an indication that the aircraft were not too far from *Intrepid*. As the time of this air activity was just as dawn broke, the two aircraft above us had most likely left mainland Argentina under cover of darkness.

Listening to their conversation Andy was not able to discover what their mission was, but he was aware from the beginning that the aircraft had come across a target of opportunity that was to be exploited – that target was HMS *Intrepid*. The Argentinian pilots had agreed to alter their course to return and overfly their new target, formulate a policy of attack, return and carry out the attack. While all this air activity was going on I had gone to the Captain's cabin to inform him of the potential threat from enemy aircraft by asking if he could hear aircraft above, to which he replied 'Yes they are ours'. When I advised him to the contrary the Captain immediately ordered the bridge to alter course in the direction of a wide area of very dense snow cloud that was to the side of our current position.

On my return to the EW cell I listened to Andy give a direct translation of the continuing conversation between the Argentine pilots; on retracing their route there was no sign of the vessel they had previously seen. I spent the next few minutes listening to Andy, then reported to the Captain; it was evident that the Argentinian aircraft had used too much fuel looking for a missed target, and as a result they would have to abort their original mission and return to mainland Argentina. Then the intercepted frequency in question became silent just as it had been a few minutes earlier. Satisfying himself that his vessel was now no longer in danger of attack, the Captain set a new course to return to anchorage in San Carlos Bay.

The majority of frequencies stored in our automated intercept equipment were those of aircraft in common use by the Argentinian Air Force. A couple of days after this incident, activity acquired through the use of this equipment resulted in a rather disturbing sequence of events. The aircraft radio transmissions on this occasion that activated the automatic intercept equipment were thought to be from a Canberra aircraft. It was known the Argentinian Air Force was equipped with these but perhaps with the exception of only one or two, most were not airworthy. This particular Canberra had been modified and was deployed in the role of a bomber against the British Land Force. It is not known how many times previously, if any, this particular aircraft had overflown the Islands but it was obvious from translated intercepts that on this occasion it was on a mission to drop its payload on a target or targets located on the Islands.

During the aircraft's communications with others externally, it was obvious that the cockpit frequently passed instructions and orders to cabin crew through the aircraft's internal communication system. How these internal communications came to be transmitted outside the aircraft will never be known, but it was during an internal communication conversation between the pilot and someone in the bomb bay that the pilot first became aware his aircraft was the target for a missile fired by a Royal Navy vessel below him. During this conversation he was looking out of the side window of the cockpit and reporting on the missile's progress to its target; from the excitement in his voice it was obvious the missile was not far away.

Suddenly, Andy Grey who had been providing the blow-by-blow commentary of this missile attack, stopped talking, took off his headphones and left the EW cell for a short time. On his return I asked what had caused him to leave; he responded by winding back the tape on which the transmission had been recorded, handing me the headphones and asking me what I could hear at the end of the transmissions. Andy translated the final words 'We have been hit,' followed by the gargling sound of the death rattle of someone dying. The recording of this activity labelled 'the Death of a Canberra' was kept until the STD's return to the UK; what became of it eventually I do not know.

12

COMMUNICATION PROCEDURES

When using two or more radio equipments communicating as a network, it is imperative to ensure that written instructions have been delivered to, and received by, subscribers before they actually commence communication with each other. These instructions should include allocating an individual identity to all subscribing stations within the network, normally in the form of a call sign; frequencies on which the transmissions are to be broadcast, plus alternatives in the event of poor atmospheric conditions; the transmission mode, whether it be voice, printer, Morse or facsimile; and times of operating schedules. As radio waves can of course be intercepted by others, consideration must always be given to the security of the transmission's content. Procedures such as the inclusion of predetermined code words and cover names to hide the true meaning or significance of an order or instruction should be used as frequently as possible. While these procedures are not in themselves 100 per cent secure, they at least go part of the way to safeguarding a great deal that is considered sensitive. They make it more difficult for unauthorised listeners to understand the content of the transmissions and can frustrate an intercept operator. For the more sensitive transmission, consideration should be given to either encryption of the content prior to transmission (offline encryption), or a secure form of transmission (online encryption).

From commencement of the first Argentinian intercept it was apparent that the networks of most importance to the STD were using fixed call signs when communicating, although on occasion more than one call sign from the same ground station was in use. All the call signs used were from no obvious base. Furthermore, the number of letters in use was not constant, with the letters in most cases forming undefined words such as 'Capanga', 'Mercedes', 'Romero', 'Aguila' and 'Gato'. In a modern Western military force, with its own integral communication units, most communications and operational traffic are usually originated by those of officer rank but the actual job of verbally communicating between the radio stations within a network is done by specialist subordinates. However, most transmission on the Argentinian networks of interest to the STD were operated by Argentinians of officer rank, with the call signs in use most probably names. The name was not only to identify the officer himself, but also the unit he commanded, as opposed to a normal radio station call sign where the allocated call sign refers to the radio station itself, and not to a particular operator. Intercepted frequencies used on these networks were fixed but changed between day and night use.

Considerable use was made of predetermined code words and cover names. Voice communications only were used. Very little operator 'chat' took place and there was an attempt to make nearly all transmissions secure by the use of offline encryption. In comparison with the average British officer communicating by radio, these Argentinian officers were extremely professional. Their skilful use of cover names and code words was of the highest standard and better than a great many 'professional operators' that I have listened to.

It is normal procedure that when lines of communication over air waves are active, consideration is given to security. Ideally, no one but the nominated recipient should even find the message. If the message is intercepted, no one should be able to read it and no one should be able to discover the identities of the sender or his/her location. In order to comply with these basic security requirements it is normal to use a form of encryption technology.

Simple encoding schemes have existed since the fifth century BC, but often a simple code or cipher can easily be cracked by those with the necessary equipment, facilities and knowledge to intercept such communications. However, in a modern electronic society cryptography has developed to encompass other features such as data integrity and authentication to deter would be 'hackers' and spies attempting to break into secure communications.

Cryptography, in a broad sense, is the study of techniques related to aspects of information security. Hence cryptography is concerned with the writing (ciphering or encoding) and deciphering (decoding or decryption) of messages in code. Without going into the theory of machine and/or computer-generated ciphers, three of the most common 'manual' systems of enciphering in use are the Polybius Square, Transposition and Substitution ciphers.

The Polybius Square – or 'Polybius checkerboard' – named after a second-century BC Greek historian, is one of the simplest tools in cryptography. Create an equal number of rows and columns to form a 'perfect square'. The smallest size of 'perfect square' should at least contain as many different squares as in the alphabet of the language in use; there is no maximum size 'perfect square' (the example below is based on the English language with the letter J omitted).

0	1	2	3	4	5
1	A	B	C	D	E
2	F	G	H	I	K
3	L	M	N	O	P
4	Q	R	S	T	U
5	V	W	X	Y	Z

To encrypt NOW IS THE TIME, using the numbers that correspond to their coordinates, in row then column order, the text changes to 333452 2443 442315 44243215. In its basic form the use of the Polybius Square is not sufficiently secure for classified information. However, the basic square can be expanded and within the intersection of each square can be a letter, number, command, instruction, phrase or punctuation. As an additional safeguard and to improve the security of the square's content, the row and column coordinates can be changed on a monthly, weekly or even daily basis. The squares could also contain more than one character or symbol; if this is so then another set of coordinates is introduced to indicate upper or lower designated square usage.

The Transposition Cipher is an encoding process that does not change any of the letters of the original message but changes the position of the letters. THE QUICK BROWN FOX becoming EHT KCIUQ NWORB XOF is the simplest example and of course easy to decode. Transposition ciphers are like jigsaw puzzles; all the pieces are present, it's just a matter of putting them in the correct order.

The Substitution Cipher is an encoding process that maintains the order of the letters in the message, but changes their identity. For example, Morse code is a substitution cipher in which each letter is replaced by a specific set of dots and dashes. Many substitution ciphers use only one alphabet and are called 'monoalphabetic', where one and only one letter is substituted for a particular letter in the message. Such a cipher scheme is easy to remember, but is also vulnerable to cracking/breaking using frequency analysis (letter counting). With a large encoded message derived using a monoalphabetic substitution cipher, it can readily be 'cracked' by comparing the frequency of letter occurrences in the coded message with the frequency of letter occurrences in the language used for the message. A well-known substitution cipher known as shift cipher or Caesar's Code was named after Julius Caesar, who is believed to have used it to send important military information. The basic principle required of a shift cipher is that a 'key' number 'k' is agreed upon by the sender and receiver, the standard alphabet is shifted 'k' positions so that the 'k'th letter in the alphabet is substituted for A, k+1 = B etc. If k=7, then the resulting shifted cipher means that A is replaced by G as shown below:

A B C D E F G H I J K L M N O P Q R S T U V W X Y Z
G H I J K L M N O P Q R S T U V W X Y Z A B C D E F

Using this cipher the message text JUMPS OVER THE LAZY DOG becomes PASVY UBKX ZNK RGFE JUM. In order to improve the security of a substitution cipher more than one alphabet can be used. Such ciphers are called 'polyalphabetic' (or Vigenere Cipher, named after Blaise de Vigenere, a sixteenth-century Frenchman) meaning that the same letter of a message can be represented by different letters. Such a multiple correspondence makes the use of frequency analysis to crack the code much more difficult. The Vigenere Cipher was considered to be practically unbreakable until 1863 when a Prussian military officer devised a method to determine the length of the keyword and then divide the message into a simpler form to which letter frequency analysis could be applied.

The three types of encryption as explained above, because they have been in use for many centuries and long before the introduction of computers and other electronically generated machines, have one thing in common: in their simplest form all can be encrypted manually. Therefore, it is reasonable to say that with a little knowledge of code breaking, all can be *decrypted* manually. For specialists employed in decryption duties an abundance of patience is required. If for example, the network being copied uses an encryption system based on Transposition Ciphers with a daily changing 'key', once in possession of the entire content of the intercepted message, to a cryptanalyst it would be like building a jigsaw puzzle; all the necessary pieces used to compile the jigsaw are in the box but the way the jigsaw is constructed changes daily. Once the message has been correctly assembled all further messages intercepted during the day will be in the same decrypted format.

Use of a Polybius Square generally takes a cryptanalyst much longer to recover the content. Unlike a Transposition Cipher where the entire message in its jumbled form is hopefully recovered, individual intercepts of messages using a Polybius Square produce limited sub-squares of the major square resulting in only partial recovery. To recover a fully encrypted square is time consuming and dependent on the frequency of use by the originator, bearing in mind the cryptanalyst is working from a blank canvas; only with a thorough knowledge of the target will he or she be in a position to piece together the content of the square. If as previously mentioned the originator uses daily changing coordinates, recovery takes longer.

Prior to the introduction of the Army trade of EW Operator, when those employed in Sigint were trained as ANSI, part of the trade training prospectus included training in the techniques of cryptography. Some intelligence analysts excelled in this specialist field and later, after advanced training, were gainfully employed as cryptanalysts. Fortunately within the STD there were several ex-ANSI, now EW Operators with varying experience in the field of cryptography. Unfortunately I am prohibited by the Official Secrets Acts from all mention of any specific activities connected with cryptanalysis from any phase in my military career. The reason for my being 'gagged', I am reliably informed, is because the sensitivities of all intelligence activities and of Sigint in all its aspects from tactical (EW) to strategic, civil and military, may long outlast that of the intelligence which they yield, and the outcomes (such as military victories) which that intelligence helps to yield. No aspect is more sensitive than that of cryptanalysis. There is no specific time limit on these sensitivities (no '30-year rule' such as is applied to other UK government secrets). Indeed some technical aspects of the cryptanalytic activities at Bletchley Park have still not been released even though the general story became public in the 1970s and the surviving decrypted German texts were released in the 1990s.

Since the British government has issued strict instructions to ensure no leakage of information relating to the area of Sigint and in particular cryptanalytic activities, to some involved in the world of Sigint it came as a bolt from the blue when on 24 June 2010, GCHQ and the National Security Agency/Central Security Services (NSA/CSS) announced they had declassified documents from the period 1940–1956, previously held under the UKUSA Agreement first signed in March 1946. Prior to this release date, those with knowledge of the existence of such an agreement knew only too well that mention of its existence outside of inner secure circles could result in serious repercussions.

The UKUSA Agreement was confirmed at government level, but the word 'agreement' rather than 'treaty' was used. This was deliberate because 'treaty' would have encouraged the involvement of politicians from both sides and this was to be avoided. It was formulated by the then British Government Code and Cipher School and the US War Department, and initially covered the production, exchange and dissemination of all special intelligence derived by cryptanalysis of the communications of the military and air forces of the Axis powers including their secret services. The Agreement called for the complete interchange of technical data and for the dissemination of the relevant data to all field commanders through special channels and subjected to special security regulations.

As a commander of a tactical communications unit, consideration should be given to outstation locations to ensure all are established in close proximity to those they are providing communications for, and as free from obstructions that may block radio transmissions as possible, so that messages need not be repeated or be sent via other network stations to the subscriber. However, in a war, choice of outstation locations is often dictated by other factors resulting in deterioration of signal strengths and the inability to communicate in some atmospheric conditions. The deployment of a communication intercept site also faces the same problems. In the case of the deployment of the STD, the unit was entirely in the hands of HMS *Intrepid's* Captain. Where he went we followed; consequently on some occasions the signal strength of the Argentinian radio communications bordered on the non-existent, and sometimes was so strong that we had to turn down the volume control to its lowest.

One very good example of this that very nearly caused an international incident was in the STD's attempts to copy and analyse a particular message that appeared to suggest the Argentinians were planning a gas attack on the British Land Force. With HMS *Intrepid* in an unhelpful location and attempting to work through extreme atmospheric conditions, we intercepted a message sent from the Argentinian Ground Forces HQ at Port Stanley that seemed to indicate a gas attack, but with only a limited amount of information available what should I do? If my initial reading of the message was correct then the Task Force Commander would have to be told immediately in order to carry out all necessary preventative measures. However, if I was reading something that was not factual, through lack of sufficient evidence, we could lose our credibility.

My course of action was to notify by signal the EW cell on HMS *Hermes* that the STD held limited evidence to suggest that the Argentinians may launch a gas attack. Within the Sigint world in those days the use of the words 'limited evidence' meant exactly that and for the recipient of the message to watch this space for any possible further developments, and not to accept the information as a categorical certainty. The EW cell on HMS *Hermes* thought differently. The next thing we knew was that instructions had been given to the entire Task Force to ensure they carried full Nuclear Biological Chemical (NBC) clothing at all times and all those wearing a full beard, mainly Royal Navy personnel, were to parade clean shaven. In the event of a gas strike and the need to wear respirators, a beard would make the respirator ill fitting and allow gas to penetrate without first going through the filter system. These orders were not taken too well, particularly as a great many men had deliberately left their NBC clothing on board ship when disembarking in the belief it would never be needed and had to return to the ships to collect it. Many of the bearded personnel objected because their wives and girlfriends had never seen them without beards and might not recognise them on their return to the UK!

After the order to carry NBC clothing, I received a message from the EW cell on HMS *Hermes* ordering me to immediately deliver to them all raw intercept material relating to the use of gas attacks because the Force Commander, Admiral Woodward, was preparing a case to go before the United Nations on the use of gas by the Argentinians. While I was more than willing to do as ordered, I also thought 'what's the use?' All I possessed, because of very bad communicating conditions, was a possibly corrupt piece of intercept that I was holding awaiting further confirmation, in the hope that the Argentinians would repeat the transmission in its entirety. I did respond as requested but only to reiterate that, as already said, the STD only held tentative evidence to suggest a gas attack. There was no further mention of it. This proved to be another example of the command structure having little in-depth knowledge of its subordinate organisations.

With the break out from the bridgehead by elements of 3 Brigade, the establishment of a rear echelon logistic platform at San Carlos and the arrival of 5 Brigade, the Argentinian Ground Forces had become more or less rooted in their initial locations around the Falkland Islands. It was obvious that if they attempted any movement by land or air the chances of reaching their destinations would be negligible; consequently their forces on West Island at Fox Bay and Port Howard were isolated, with their only form of communication being radio. It is not known what prompted the controlling authority of the Argentinian Land Forces' main command net to broadcast a message containing changes and amendments to their signals operating procedures, other than the fact that because they could not physically deliver orders, they had no alternative but to resort to the radio. The message I refer to was by size, the largest sent to its subordinate outstations throughout the entire period of hostilities. It also took the longest amount of time by far before all subscribers to the network had received the message in full because of the appalling atmospheric conditions on the afternoon that it was transmitted.

Fortunately, because of where HMS *Intrepid* was located at the time, the STD had little difficulty in intercepting this critical message, and because of repeat after repeat to enable the Argentinian outstations to receive the message in its entirety, we could almost guarantee we had a perfect copy of what the controlling authorities had sent. Whereas prior to the intercept of this message various codenames, nicknames and cover words had been used, with their meanings never apparent to the STD, this message contained names and words, some fixed, others one-time use only, to disguise the four cardinal points north, south, east and west, the seven days of the week, to activate the change over from day transmitting frequencies to night frequencies and many other procedures used to safeguard the net security. From the intercept of this message and throughout the remaining period of hostilities, the Argentinians could do very little to prevent us from hearing their proposed intentions once orders or instructions were transmitted over the command net.

After the successful advance on Goose Green, Argentinian supplies of napalm were discovered near the air strip at Goose Green. This terrifying weapon takes its name from the original ingredients of coprecipitated aluminium, naphthenic and palmatic acids which, when mixed with gasoline, produce a thick jelly-like substance for use as a sticky incendiary gel, that when dispersed over a target sticks to the surface and generates sufficient heat to cause ignition of the surfaces themselves. Anyone caught in the open by napalm would be splashed with the burning fluid, surrounded by flames and deprived of air, so their chances of survival were slender. In addition to knowing that the Argentinians had a supply of napalm, I was not totally surprised by the fact that several weeks after the incident of the message regarding a possible gas attack, I was shown photographs taken at Port Stanley airfield of various sizes of gas canisters stockpiled outside a building. The canisters had letters and figures stencilled on them and I was asked if it was possible to identify the contents of the canisters from their markings. Unfortunately the STD was unable to give any clues as to the identity of the contents. In retrospect because of the discovery of napalm and the gas canisters, I now feel more than justified that I informed the EW cell on HMS *Hermes* of the possibility of a gas attack on the Land Task Force.

The mention of gas and napalm reminds me of another incident in my career involving toxic substances. On 26 April 1986, a reactor at the Chernobyl nuclear power plant in the Soviet Union exploded, sending a plume of highly radioactive fallout into the atmosphere over an extensive area including Western Europe. The amount of radioactive fallout from this accident was almost 400 times greater than that from the atomic bombing of Hiroshima. At the time I was serving as the Senior Intelligence Officer with a signals regiment just inside West Germany, very near the West German border with Holland. The operation centre where I worked was a cavernous open plan room, most of which was built below ground level. The room had no natural light and air was distributed via large turbines on the roof of the building. I remember that after the accident considerable concern was expressed in the press for the Welsh hill farmers, because some time after the accident their sheep had eaten contaminated grass and a great many had died. Meanwhile, those several hundreds of miles closer continued as if nothing had happened. That was until the spring of 1987, when my unit was informed that scientists were to visit the unit to carry out an inspection of the air filtration units on the roof of the operation centre. This was done by men clothed from head to toe in protective overalls and breathing gear. Tests were carried out after which the old filters were removed from the roof, placed in sealed containers and taken away. We were never informed why the filtration plant should have been of great concern or what became of the filters. However, it was believed by all that greater concern had been given to the sheep on the Welsh hills than those much closer to the epicentre of a nuclear disaster.

13

THE PLANNED COUNTER ATTACK

During the first two weeks after the Task Force's arrival, other than the penetration patrols by the SAS and the naval fire on the radar site located within a short distance of Fox Bay, no other hostile actions had been carried out by the British against the Argentinians on West Island. The Argentinian logistical chain had been broken, resulting in their inability to re-supply their troops on the ground. The intelligence gleaned from the contents of their radio messages suggested that morale was deteriorating through the limited supply of food and boredom brought about by a lack of activity.

On occasions it was apparent that instructions were passed to the commands of the ground forces located on West Island from the commanders at the Argentinian HQ in Stanley for all ranks to write letters to their families and friends. The instructions included guidance as to what should be written, which was along the lines that the war was going well, in favour of the Argentinians, and that the troops were comfortable and well fed. Having written the letters the young troops on the ground would have no idea that they couldn't be delivered to their families and friends because they were not aware of their isolation. It was obvious to us that our enemies were not well fed, because the STD had intercepted messages between commands mentioning that some of the Argentinian troops were suffering from malnutrition, which in at least five cases led to the deaths of young men who were believed to have been buried at first in the church graveyard at Port Howard. I do not know what became of the deceased after the end of hostilities.

With the safety and security of the crew and the ship itself utmost in her Captain's mind, it was by now customary for HMS *Intrepid* to return to her anchorage in San Carlos Bay by sunrise every day. With departure of 5 Brigade and 3 Brigade to establish new forward areas, the area in the vicinity of the bridgehead at San Carlos Bay, with the exception of those few hours around midday when the Argentinian Air Force made their daily aerial attacks, appeared much more calm and quiet.

In the bay there was a constant unloading of supply ships with their cargos transferred to temporary store before being moved to the forward areas. The field hospital was kept busy treating both British and Argentinian casualties and a temporary PoW camp was established, although activity here was limited by insufficient numbers of Spanish linguists. With the majority of the Land Force some distance from the bay, the only British troops left at the 'rear' were a few specialist logistic and medical personnel and a depleted 40 Commando; at least half of their strength was sent forward to make good the losses from the Guards regiments as a result of the bombing of *Sir Galahad*.

The more that HMS *Intrepid* returned to her anchorage, the more I got to thinking how vulnerable we would be, located in the area far west of East Island, to an Argentinian counterattack, particularly if it were launched from East Island. None of the British commanders appeared to be overtly worried, otherwise they would not have left their rear echelon in such a vulnerable situation. However, it was not long before my worries were realised. Intercepts of traffic between the Argentinian HQ at Port Stanley and the Command HQ Fox Bay, suggested that a ship whose identity escapes me but I strongly suspect it may have been the ARA *Bahia Buen Suceso*, previously thought to have been damaged and beached not far from Fox Bay after being attacked by a British Harrier jet, had been salvaged by the Argentinians and made seaworthy. Shortly after this, what I thought was a very significant piece of information had been passed to the Task Force HQ staff. A radio communication between the Argentinian HQ at Port Stanley and the Command HQ Port Howard was intercepted discussing the possibility of sending reinforcements of 500 airborne troops north of Port Howard under cover of darkness. All information relating to a possible Argentinian counterattack was treated as the STD's top priority, and once intercepts relating to this activity had been analysed to the best of our ability, reports were immediately forwarded to 3 Brigade HQ, in particular the intelligence section supporting the HQs.

To our dismay the contents of later messages we received from the HQ staff on HMS *Fearless* suggested that there was little interest in intelligence that indicated the Argentinians were contemplating reinforcing West Island, or even the possibility of an Argentinian counterattack. During my now infrequent visits to the EW cell on *Fearless*, I was quietly informed that perhaps we had made an error of judgement in the interpretation of the limited material available concerning West Island, particularly in the Argentinian garrisons at Fox Bay and Port Howard. With the knowledge that no organisation other than the STD had any evidence about what was happening on West Island, I treated this rebuke with the contempt it deserved and carried on working in the only way I knew how.

As the days passed it became increasingly evident that time spent by the Argentinians on communications between the controlling authority in Port Stanley and the command HQs at Port Howard and Fox Bay had increased considerably and it eventually became evident during the STD's processing of this information that the Argentinians were definitely planning a counterattack against the British Land Forces around San Carlos Bay. To do this they expected to transport their troops in the Fox Bay area by using the repaired and now seaworthy *Bahia Buen Suceso* as a ferry, crossing to East Island to an area just south of San Carlos. The troops located to the north at Port Howard, together with the airborne troops, would be transported in helicopters that had remained hidden from the British Land Force, across to East Island to an area north of San Carlos. Having taken up their new positions, the Argentinian troops would then advance to positions north and south of the original bridgehead at San Carlos Bay.

In reality a counterattack, if it had been carried out, would have been a walkover for the Argentine Force. From very early on in the period of hostilities the STD, through its analyses and processing of the Argentinian Forces' communications networks, was able to ascertain that the combined strength of Argentinian forces in and around Port Howard totalled in excess of 2,000 and the enemy strength in and around Fox Bay, was approximately 1,500. These unit strengths had been regularly updated and reported back to the intelligence cell on HMS *Fearless* and to sponsor branches in the UK. The combined total of Argentinian Ground Forces, including the airborne troops landing in the vicinity of San Carlos, would have been approximately 4,000. The token reserve British Land Force of approximately 200 would have been insufficient to defend the food, ammunition and fuel storage sites at San Carlos, as well as the imprisoned Argentinian PoWs. If the

Argentinians were successful in their advance on San Carlos, added to the forces from West Island would be approximately 2,000 Argentinian PoWs, to give the Argentinians a combined force in excess of 6,000 troops to the rear of the British Land Force, with approximately 8,000 troops forward in and around Port Stanley. In addition to the Argentinian superiority in number of troops on the ground, they would also have had access to the British stores of food, ammunition, medical supplies and fuel.

As the days passed it became clear that the counterattack, the plans having been formulated and agreed by all concerned, was to be executed at 0800 on 15 June. However, the STD was aware through Argentinian intercepted traffic that the plans to include the 500 airborne personnel could be in danger of cancellation due to bad weather. As it eventually turned out, all the Argentinian plans never came to fruition. As history reveals, the Argentinians agreed to a ceasefire and eventual surrender at 2359 hours on 14 June, eight hours before their planned counterattack on West Island. If it had been successfully executed and the Argentinian force on East Island had increased by 6,000 troops, I would assume the eventual outcome of cessation of hostilities would have been a little different, and perhaps the British Task Force would not have returned victorious.

14

THE SURRENDER

Late morning of 14 June 1982, and the British Land Force advance towards Port Stanley was progressing well. At this time I was in the 'Standby HQ' on HMS *Intrepid* where the topic of conversation was that the Commander of the British Land Forces was not too happy with elements of the Paras. Rumour Control suggested that the OC A Coy, 2 Para, on reaching and securing his objective virtually unopposed, had continued to advance in the direction of Port Stanley with one platoon of troops. As they did so, Argentinian troops came out of their fox holes waving anything white they could lay their hands on and retreated in front of the Para platoon in the direction of Port Stanley. When the Commander British Land Forces was made aware of the OC A Coy's activities, he was not amused. He wanted to know why the officer had ignored his instructions to advance towards his objective and, when secure, hold his ground until further orders. Accepting that on this occasion, by using his initiative, the officer had made the right decision, he ordered the Para platoon to continue advancing towards Port Stanley. But, immediately they met any enemy opposition or came under fire, they were to hold their ground or return to their original objectives.

Several hours later it was rumoured that the OC A Coy had contacted HMS *Fearless* by radio to state that because the Argentinian troops in front of him were heading in the direction of the Port Stanley airfield, they would soon run out of land because the it was on a small peninsula surrounded on three sides by the Atlantic Ocean. Therefore he was going to remain at his current location and allow the Argentinian troops to continue at their own pace towards the airfield. It was evident to all at this time that the Argentinian Land Force had lost the will to continue and surrender was inevitable.

At 9pm a surrender document was signed by Major General Moore on behalf of the British government and Major General Mario Benjamin Menendez, governor of Islas Malvinas on behalf of the Argentinian government. Whilst Major General Menendez had surrendered the entire Argentinian Force deployed on Falkland Islands, he initially stated that the Argentinian troops located on West Island were technically a separate garrison and not under his command. Consequently, he had no authority to surrender it. In response the British stated that West Island was geographically a part of the Falkland Group and therefore under the jurisdiction of General Menendez.

The question of responsibility for West Island eventually resulted in one surrender document being signed on the 14th and a second document signed by the Commander Argentinian Forces on West Island, who by this time had also agreed with General Menendez to the surrender of his troops. The task of accepting the surrender of the Argentinian troops on West Island fell to Lieutenant Colonel Malcolm Hunt, Commanding Officer of 40 Commando, with the surrender and signing of the documents to take place on the morning of 15 June. At about the time the second set of surrender documents were being signed, the proposed Argentinian counterattack, as mentioned in the previous chapter, had been scheduled to commence.

The ending of hostilities was sudden, perhaps just as well for the British Land Force. The advance on Stanley had been accomplished extremely quickly but at the expense of stretching the logistics chain. Keeping the forward elements of the force supplied with essential stores and ammunition was becoming very difficult through lack of transportation, mainly helicopters, and the distance to be covered over ground. Consequently, by 14 June the forward troops were almost completely out of ammunition and had to resort to close quarter combat with fixed bayonets. Had the Argentinian High Command known the problems of the British logistical support, I have often wondered if they would have been so quick to surrender.

In the late evening I was aware that a junior British naval officer had been placed on standby to offer linguistic support during the signing of surrender documents at Port Howard the following day. Having some knowledge of this particular officer I was aware, through no fault of his own, that his linguistic skills were of insufficient standard to represent the British government at such an important event. Therefore, I informed the Captain of HMS *Intrepid* of my views and suggested to him that WO2 Andy Grey should be in attendance. This was agreed and I duly informed Andy that he would be picked up by helicopter early in the morning, taken to San Carlos Bay to pick up Lieutenant Colonel Hunt and accompany him to Port Howard for the signing of the documents. Andy's response was that he couldn't possibly carry out linguistic support if required to do so because he was not of commissioned rank. From his experience on Spanish translation duties with Latin American army officers, it was evident that no notice would have been taken of what he said because he was not a commissioned officer. I informed him that the problem would be quickly rectified and asked the Captain of *Intrepid* to approve a field commission for the day; this he agreed. I then returned to Andy, gave him my badges of rank and told him to be on parade in the morning to carry out linguistic duties. Before he left, I gave Andy my camera with a request to photograph anything he thought might be of interest. Unfortunately, he had very little time to take many photographs but among the few he did take was of the Union Jack flown on West Island immediately the surrender documents were signed.

On his return to *Intrepid* I inquired how his morning had gone. Andy had mixed feelings. On their arrival at Port Howard while hovering to find a suitable place to land the helicopter, he noticed several hundred Argentinian soldiers seated in a clearing close to a building the Argentinians had used for their HQ, which was also close to the area the pilot had decided to land his helicopter. While the helicopter was hovering, all Andy said he could think of was what if an Argentinian decided he didn't want to surrender and fired on the small British contingent?

Having alighted from the aircraft the British group were taken inside the HQ building where they were introduced to an Argentinian Major General and other senior officers. The Major General asked if he could make a small presentation before they got down to the official business of signing the surrender documents, which was agreed to. Silence descended on the room when an Argentinian officer walked in wearing full uniform, and from his insignia Andy quickly realised that he was a member of the Argentinian Special Forces. This officer was carrying a small cushion on which was a Special Forces beret and cap badge. The Major General then proceeded to make a speech in which he praised the bravery, courage and strength of character of Captain John Hamilton, SAS who was killed by

the Argentinian Forces (see Chapter 9). He said that no matter what they
threw at him, Captain Hamilton would not go down; he remained standing
until all his ammunition was used and only when he thought his Corporal
had escaped did he decide to lie down. The presentation of the beret and
cap badge was a small token of their acknowledgement of having been up
against an extremely courageous and professional soldier and was, if possible, to be handed to his next of kin.

Andy then said that all present wanted to get on with the signing of the
document but there were several important things yet to be done, mainly
revealing the locations of booby traps and other explosive devices. This was
met with complete silence from the Argentinian officers present, so he asked
to speak with an engineer officer. A sapper officer was found and brought
to the room. This officer initially refused to admit that the Argentinians had
laid mines and booby traps. With that Andy asked him where the ration
store was and could he be taken to it?

The officer duly escorted Andy to a large shed but when Andy asked to
be allowed entry this was refused because the shed and surrounding area
had been booby trapped. Andy then asked the Argentinian to accompany
him to the jetty; when they arrived Andy suggested the officer should walk
along to the end, the officer again refused because the jetty was also booby
trapped. Andy then requested a map of the local area showing places that
were mined or booby trapped; the officer said they did not have an official
map of the area showing what was required but he would get a map and
mark any dangerous areas on it from memory. Most of the booby traps were
simply grenades with the secure pin removed and gently placed on the
ground; any movement close by would cause the grenade to explode.

The officer was impressed with Andy's knowledge about the ration shed
and jetty and asked how he knew. Andy of course refused to comment;
he wasn't going to tell the Argentinian that the STD had been intercepting all their radio traffic and frequently intercepted messages indicated that
Argentinian troops had been stealing food from the ration store and also
that they feared British frogmen would land on the Port Howard jetty, so
they had decided to booby trap both places.

On completion of the signing it was arranged for HMS *Intrepid* to transport
the PoWs from Port Howard to San Carlos Bay, prior to their repatriation back
to Argentina. Once the surrender documents had been signed, *Intrepid* left her
anchorage at San Carlos Bay, headed across the Falkland Sound and tied up
alongside a safe jetty at Port Howard to commence the boarding of the PoWs.

Not long after the boarding began, I asked Andy to go below decks and return with a report on how things were progressing. He returned to say that the Argentinians were lined up in single columns; approximately every third man had a 7lb tin of corned beef under his arm, which represented their sole ration stock. As they approached the ship they were relieved of their weapons, which were placed in piles on the dockside, then they entered the hold compartment of the ship, which had been previously cleared of all British vehicles and equipment. I asked Andy who these Argentinian troops were and if he had any idea from which military units they came. Andy didn't known but said he would endeavour to find out.

Andy was gone for a considerable time and on his return he said that having made inquiries of several British officers as to identities of the Argentinian troops and getting no answers, he had better take things into his own hands. By this time there were several hundred Argentinian troops on *Intrepid*, and having found a table, a blank exercise book and a couple of ballpoint pens, Andy stood on the table, asked for silence and welcomed all the troops on board. He then asked if there were any officers on board, to which he got no reply. He then said that in a British officers mess it was customary for visiting officers to sign the visitors' book; therefore, as the 'hold' was to be considered an 'officers mess' for an unknown period of time, would the officers kindly come forward and sign the visitors' book?

The first officer to sign the improvised visitors' book was a padre; he asked Andy what he should put in the book and Andy replied, name, number, rank, specialist appointment and parent unit. After the man completed his details he was followed by all the other officers. Andy then handed me the book.

Looking at the officer appointments, I noticed that one had signed in as the regimental records officer. When I mentioned to Andy that this officer might be a candidate to ask a few questions, Andy said that he remembered who he was because he had brought on board a very expensive looking leather Gladstone bag. I then asked Andy if he could return to the hold and temporarily confiscate the briefcase. Andy returned with it and when I opened it the first item we saw was a pistol, complete with several clips of ammunition. So much for the weapons checks!

The owner of the briefcase was in fact the unit records officer and in his bag was a complete history of the General O'Higgins Regiment – at the time I was slow to comprehend that the O'Higgins Regiment was believed to be located at Fox Bay, as part of the Argentinian 8th Regiment and not at

Port Howard as part of the 5th Regiment – together with names and personal details of those currently serving with the regiment. In addition to the records was a piece of paper on which handwritten notes had been made. This was not classified but when Andy started to transcribe its content from Spanish to English, it was obvious that what we were looking at were notes taken at a meeting the officer attended on the planned counterattack scheduled for 15 June. The translated version was a virtual copy of what we had included in our reports to British Land Force HQ in the Falkland Islands and sponsor branches in the UK.

What eventually became of the briefcase and its contents I do not know; I made arrangements for it to be forwarded to the intelligence cell on HMS *Fearless*. I did hear much later that it had eventually turned up at the Intelligence Corps Centre in Ashford, Kent. I retained the copy of the handwritten notes of the possible counterattack in the event that I might have at some time to defend my reporting of such a possibility. A copy of this document is included as Appendix 2; as I have already said it bears no security classification markings and to those not in a position to know better, would have appeared to be an innocuous piece of paper. In reality, for the STD it represented one of the 'Jewels in the Crown' that contributed so much to the detachment's operational success during Operation *Corporate*.

On completion of the transfer of Argentinian PoWs from Port Howard to San Carlos Bay, the Captain of *Intrepid* called me to the bridge where he informed me that *Intrepid* would not relocate to an anchorage off Port Stanley where the rest of the fleet was gathering prior to returning to the UK. This was because the British authorities had agreed that while the repatriation of the Argentinian PoWs was in progress, the Argentinian Command would be allowed to maintain radio communications with the mainland in order for the Red Cross – who were overseeing the repatriation – to communicate with their officials in Argentina and the Falkland Islands. While this was going on, *Intrepid* was to remain in a much reduced state of operational readiness and we were to carry out the interception of communications between the Argentinian Ground Forces on the Falkland Islands and the Junta headquarters in the Argentine. Although the Red Cross radio link was active 24 hours a day, nothing of intelligence value was intercepted. The very few occasions it became active were to report the departures and arrivals of ships carrying the returning Argentinian troops.

After about a week of listening to this repatriation communications link, the Captain asked if I could foresee a date when the link would be closed down. He was aware that the fleet was almost ready to return to home shores and he considered it would be a great morale booster for the ship's crew if they were to return to UK with them. I learned shortly after my conversation with the Captain that the repatriation of the Argentinian Force had been successfully completed and therefore there was no need for the STD to remain on board. However, I did ask the Captain to do us a favour and put us ashore at Port Stanley, which he kindly agreed to do. So after almost nine weeks at sea, with our radio equipment safely reinstalled in the racks of our Land Rovers, we said our last farewell to friends and colleagues and finally set foot on the Falkland Islands for the first time.

Why didn't the STD remain on HMS *Intrepid* and return to the UK as part of the embarked crew? The reason was that our sponsor branches in the UK had decided that if the hostilities ended favourably for the British, then it might be of some use to create a semi-permanent unit on the Falkland Islands in order to monitor the future activities of the Argentinians. Therefore, until a definite decision had been made, we were to remain in situ.

15

FORMATION OF A PERMANENT SIGNAL UNIT

It was mid morning, still comparatively dark for the time of day and everything we touched was soaking from the previous night's torrential rain. The crew of the STD were out on the open deck finalising preparations for our disembarkation in HMS *Intrepid's* LCU to shore at Port Stanley. It was from here that we caught our first sight of Port Stanley, and as capital cities go it was tiny; what stood out most were the assorted colours of paint the islanders had used to decorate the corrugated iron roofs of their homes. As we sailed closer to land it was obvious that although the population of the Falkland Islands was small, there was plenty of land. However, building land appeared to be at a premium because the majority of the houses we could see were rather small and compact with not much space between them. With the detachment's radio equipment once again installed in its original racking, the two Land Rovers were loaded onto two of *Intrepid's* LCUs and the STD set off towards the town's harbour.

During this short journey my men travelled in comparative silence, the air of melancholy was very unusual, especially as previously everyone had been so upbeat, happy the war was over and pleased to have survived. The reason for this was that none of them were looking forward to a further indefinite period of time in the South Atlantic. They had come to fight a war and now that it was over they, along with the entire Task Force, wanted to return to the UK.

Prior to our departure from HMS *Intrepid*, the Brigade EW Liaison Officer, Major Mike Dawkings, had informed me by signal that the STD was initially to establish a strategic operations cell in Government House, The Governor, Rex Hunt, had been removed from office by the Argentinians some considerable time prior to the arrival of the British Land Force. In addition to being informed where we were to be temporarily established, it was mentioned that the detachment was also to have an increase in manpower, in particular Spanish linguists who had been previously employed on other ships in the Task Force.

With our feet on land after almost nine weeks at sea we set off in the direction of Government House. Two of the detachment's crew drove the vehicles while the rest chose to walk. The nominated vehicle drivers considered this unfair so we made a couple of pit stops to allow the drivers to change and everyone to get a little exercise after being so long cooped up on a warship. Following the beach road from where the LCUs dropped us off, we started to see places and things that had often been mentioned in the Argentinian radio communications we had intercepted. We passed the entrance to the Stanley airport where by this time, apart from a couple of flimsy looking buildings, a radar control tower and a wind sock, the main sights were the huge piles of small arms once carried by the Argentinian troops, now forming rusting heaps by the side of the road. Moving on we passed the Upland Goose Hotel, Stanley's only shop, owned by the Falkland Island Trading Company, the offices of Lineas Aereas del Estado (LADE), an Argentinian passenger and transport airline company that flew aircraft on non-economic commercial routes, and the town hall, which at the time of our arrival was a scene of frenetic activity involving both British and Argentinian troops in a joint clean-up operation.

The initial impression on entering Port Stanley town centre for the first time was not so much created by what we saw but what we smelt; the Argentinian Forces had used all the vacant buildings as public toilets, and although most of the mess had been cleaned up, the smell of human waste had entered the fabric of the buildings and this took a long time to fade. Passing the war memorial and cathedral we could see the racecourse that had been mentioned regularly during our intercepts, and where on a couple of occasions the EW cell on HMS *Hermes* was notified of targets of opportunity, mainly helicopters, on the ground.

On arrival at Government House I was directed to the radio room as was, prior to the Argentine invasion. The radio room, while entered from within the house itself, was a much later addition and from the outside looked like a rather large shed attached to the side of the building. It was windowless and the only door was fitted with secure devices to prevent all and sundry from entering.

While inspecting the room from the outside it was obvious the Argentinians had used it for its intended purpose because feeders from several very large antennas, undamaged and all sighted towards mainland Argentina, terminated at a junction box attached to the external wall of the building. A quick test at the terminal of these aerials proved they were still operational, so, what was good for the Argentinian operators would also be good for the STD. The room itself was a tip; the radio sets used by the Argentinian Land Force had been removed, taken outside and smashed beyond all recognition, not by the Argentinians but by the advancing British forces. The floor was littered with paper, mostly ripped from books that were originally taken from the house's library; amongst the debris I noted several pages torn from reports on oil exploration in and around Antarctica and copies of Shackleton's diaries and logs from his expedition to Antarctica during 1914–1917. Unfortunately the books, diaries and logs were in too bad a state to be salvaged and were included with the rest of the rubbish for disposal by burning.

Within a couple of hours of our arrival at Port Stanley, our radio equipment was removed from the Land Rovers and installed in the radio room of Government House. After a very short break the STD was back in the business of intercepting radio communications, this time via antennas provided by the Argentinians. Shortly after establishing ourselves our own 'Mr Fixit', Alan Newman, returned with his arms full of goodies. He had been bartering with the islanders exchanging food, in particular eggs and bread, for alcohol, but his best acquisition was the use of a washing machine so we could at last wash our clothes. This came courtesy of the Governor's housekeeper; she must have thought that if we were using the house, we might as well be clean while doing so.

After establishing its operational cell, it was time for the STD to look for accommodation for itself. From my own experience of a two-year secondment to the Canadian Forces during 1971–1972, I knew that sleeping under canvas in winter without acceptable Arctic clothing and equipment was out of the question. During that tour of duty, all service personnel had to attend a 'Winter Indoctrination Course' every second year. The week-long course, held in the wilds of northern Ontario in the winter months, was to ensure that service personnel were able to operate in extreme winter weather conditions. The entire course was held in the open air without access to permanent buildings, which meant that sleeping arrangements for the students was under canvas, in an 18-panel bell tent, one man to each panel, heads on the outside and feet towards the tent pole in the middle. A petroleum heater was permitted inside the tent during the night on the condition that one man stayed awake all the time the heater was in use. We slept on airbeds which had an extra piece like an open pillow attached to one end; to inflate the bed, air was trapped in the 'pillow', which was then squeezed into the bed. They could not be blown up by mouth because this would lead to a build up of water vapour, which would then freeze. Our issued sleeping bags were designed to be used in sub zero temperatures, as was all our external clothing.

Another experience of Arctic conditions was the three weeks I spent a couple of hundred miles south of the North Pole at a place called Alert, on Ellesmere Island, situated 82 30N 62 20W which at the time was the most northerly permanently habitable place in the world. I was there during the winter, and consequently did not see daylight for the entire three weeks. There were buildings with central heating, but if you ventured outside you faced the full force of an Arctic winter. My most vivid recollection of life in Alert was when visiting the officers' mess bar for the first time. While I was waiting to order a drink at the bar a customer asked for a gin and tonic; the barman filled an ice cube tray with water, opened the nearest window and placed it on the window ledge. Closing the window he proceeded to pour the gin and tonic into a glass and to add a slice of lemon, then opened the window and from the tray, which had only been sitting outside for a few moments, tipped out perfect ice cubes. It really is like that in the Arctic Circle in winter.

As the idea of living under canvas had been rejected due to lack of suitable clothing and sleeping bags, the alternative accommodation was on board the logistic landing ship RFA *Sir Bedivere* which was berthed in Port Stanley harbour. This was our preference; at least on board ship we had central heating, hot water, showers and proper dining facilities. I shared a four-man cabin with officers from the intelligence cell at Brigade HQ, and the other members of the STD had cabins elsewhere on the ship.

The second day after our arrival at Port Stanley, I was asked to close down the radio link between the Argentinian Ground Forces on the Falklands and the Argentine Junta Command on mainland Argentina, which had been allowed to remain active at the request of the Red Cross, because by now the majority of the Argentinian Forces had been repatriated to the mainland and there was little sense in allowing it to remain operational. What the HQ staff officers at Stanley were thinking when they instructed me to turn the link off I will never know, but my instructions were for me together with WO1 Alan Newman to meet up with a small contingent of Military Police and armed Marines outside a house used as Brigadier Oscar Joffre's HQ. I didn't know if, with my armed support, I was supposed to storm the building or what. Anyway, whatever HQ might have thought would happen didn't happen because having established that, like the radio room at Government House, all antennas serving the radio station terminated in an identical junction box on the side wall of the HQ, all we did was disconnect the aerials. Having broken the radio link we then went into the house and found a considerable amount of radio equipment in operational readiness installed in what most probably would have been one of the bedrooms.

The radio equipment was unfamiliar as it was a combination of American and French manufacture; however, on the radio that was obviously in use prior to disconnecting the aerials I noticed a locked compartment and asked the Argentinian operator for the key. This request was refused. When I suggested that the operator and I should go outside with a burly Royal Marine accompanying us, the man quickly handed it over. On opening the compartment it was obvious that this was where the encryption device was housed.

Mainly to satisfy my own curiosity and perhaps learn more about the encryption device and to have the equipment included in our database, I asked the officer in charge of the armed support unit that when the contents of the house were removed that the radio equipment be handed over to the STD at Government House. He agreed to pass on my request. After a couple of days the radio equipment had not arrived, and on inquiring as to the delay I discovered that it had been taken to a dump and destroyed along with the rest of the house's contents.

In addition to attempting to get the radio equipment belonging to Brigadier Joffre's HQ, I also asked the Land Force HQ staff to inform me where I could locate any captured Argentinian radio equipment, as frequency settings and other technical parameters might make useful additions to our database of Argentinian radio equipment. My request was granted and eventually I was given several locations where captured radio installations and equipment could be found. However, on inspection of the equipments at these sites, nothing of significant detail could be found because every single piece of equipment had been completely smashed to pieces by the advancing British Land Forces, obliterating any useful information. The only items of interest left intact were a couple of VHF handheld radios of the type used by the Argentinians during the Battle for Goose Green.

I could see the potential of these handheld radios as possible props in the event of being asked to talk about my experiences of the Falklands War or as a gift to the Royal Signals Museum, so two of these radios were eventually despatched to the STD's home location in England. On my return I was informed by the unit 2i/c that being aware I had despatched a package from the Falkland Islands, on its arrival, although addressed to me personally, he had instructed the parcel be opened and not being aware of how the hand-held radios came into my possession decided they were trophies of war and confiscated them. I explained that I had been authorised to take the radios and asked could I have them back? He replied that during my absence they had been disposed of and it was too late to return them. Disposed of to where I will never know; no doubt to the same place as the other items, mementos and souvenirs in the same parcel.

One thing that was returned was the STD's log of events from the time we left Portland docks to its arrival at Government House, Falkland Islands. The log contained a daily summary of events both administrative and oper-ational written in a hardback lined book with every page numbered and the contents classified Top Secret Codeword. Because of its security clas-sification the Log had to be included in the Top Secret classified register and stored under appropriate conditions by the unit's Top Secret Custodian. The last time I saw the STD operations log of the Falkland Islands was in the summer of 1983. It would be very gratifying to know what eventually became of this record of a little piece of military history concerning the introduction of electronic warfare in the British Army.

With the closure of Brigadier Joffre's radio station the operational duties of the STD in providing EW support initially to 3 Brigade then HQ British Land Forces had been completed. Instructions had been received from our sponsor branch at DI24(A) to establish a permanent signals unit on the Falkland Islands, firstly using the human resources and equipment originally belonging to the STD, with additional manpower and equip-ment to be sent as soon as possible. With no integral communication facilities of its own this new permanent signals unit would have no means of direct communication with the UK, therefore it was provisionally arranged for all signal traffic to be handled by the satellite communication troop from 30 Signal Regiment.

This troop had been active in the Falklands providing communication support to both 3 and 5 Brigades throughout the hostilities. On reflection, in the space of about six months, the operational mode of the STD had gone from that of an air portable unit with communications provided by the RAF to a sea platform with communications provided by the Royal Navy and now to a land-based joint service unit with communications provided by the Army; I believe its called adaptability!

After approximately ten days of operating from the radio room at Government House, the new signals unit relocated to join the HQ British Forces, Falkland Islands, in accommodation located a little farther out of Port Stanley. The accommodation building had been a school and the signals unit was allocated a single room, unfurnished except for two built-in wardrobes, with no tables or racking; how were we to create a working area? Our radio equipment had to be installed inside because it was far too cold to work from soft-skinned vehicles without added protection from the elements.

The distance between the school and our living quarters on *Sir Bedivere* was such that, with no transport at our disposal, to walk to and fro would have taken more time than an acceptable lunch break would allow and because the signals unit was classified as a 24-hour shift unit, a supper meal for those on duty would need to be provided, which would entail the kitchen staff providing a cook to prepare a late meal. The simplest solution was to issue the unit with ten-man ration packs to eat in situ rather than waste time walking to the mess hall on *Sir Bedivere*.

Initially, not knowing how many ration packs would be used in the course of a week due to constantly changing unit strength and meal requirements, and because we had to order ration packs in advance, we ended up requesting far more than we needed. This was fortuitous as the ration packs, when stacked in small piles, made a good base onto which were placed the built-in wardrobe doors, creating four platforms on which we put the radio equipment; more ration packs made good chairs. The problem with this furniture was that the packs were only issued on a weekly basis; therefore as the food was eaten the seats got lower and lower, as did our work benches. Eventually the unit was issued with tables and chairs, which was just as well because by this time someone had discovered that our weekly issue of ration packs was far in excess of our entitlement and we had to use the surplus before requesting a re-supply.

On one particular occasion when I and Major Mike Dawkins were on the long walk to the *Sir Bedivere* the weather conditions were atrocious. After several days of snow, followed by a period of intense cold accompanied by sleet and rain, there was a considerable thickness of ice covering all surfaces including the roads, making walking extremely hazardous. Having great difficulty in remaining upright we eventually reached the racecourse, which was about the halfway point of our journey, when we noticed a Chinook helicopter approaching the landing pad. As the Chinook came closer both I and Mike had even greater difficulty in remaining upright because of the down draft from the blades; eventually both of us were knocked down. To this day I maintain that we must have been observed by the crew of the Chinook, as we stood out a mile in the white landscape, dressed as we were in khaki. However, they continued to land the helicopter only some 20 metres from us, so that we were knocked flat and blown across the ice in the direction of the South Atlantic Ocean. Eventually Mike managed to direct himself towards a parked car and was able to grab hold of a wheel. I continued to be blown until I hit a small fence. The following day we reported this incident to the Air HQ but nothing further was heard. Many years later it was reported that an Allied soldier serving in Afghanistan was killed when the down draft of a helicopter's blades caused nearby ground mines to explode. If our complaint had been taken more seriously then it might have gone some way to initiating the changing of Chinook landing procedures.

In a relatively short period of time the crew of the STD had settled into a routine, those on the day shift working extended hours partly because there was little else to do. In the absence of barrack accommodation with its messes and other facilities, our social life was non-existent. While the wildlife and ecology of the Falkland Islands had much to offer, we could not explore because of the restrictions on movement outside the immediate vicinity of Port Stanley. The Argentinian troops had saturated their areas of occupation with land mines and booby traps, and although some troops had been held back from the main repatriation to assist in mine clearance duties, the British engineers quickly came to the conclusion that it would take several years to clear them all, and that it might be impossible to locate and defuse all the mines, which would mean that great areas of the islands would never be walked on again, or only by sheep.

As expected there was little target operational activity. With the return of its defeated armed forces the Junta in Argentina was seeking ways to protect its own future rather than thinking of a possible return to the Falkland Islands. Life was slowly calming down. However, life for the islanders would never return to how it had been before the invasion, as they had relied on Argentina for much of their food, clothing and building materials. Some things improved, as before the war there had only been the small Argentinian passenger and limited freight airline company (LADE) working out of Stanley airfield to Argentina, from where islanders could change planes for further international travel. Their only other option had been the twice-yearly ship from the UK. Now aircraft and ships were arriving almost daily, bringing with them a variety of perishable items, day-old newspapers, and videos of British television programmes to be broadcast locally over a system similar to that used to broadcast to the British Forces in Germany.

These improvements meant that I was able to read fairly recent daily newspapers and copies of early papers that had been held up in the supply chain from Ascension Island. The older papers, mainly copies of the *Daily Mirror* and the *Sun* were full of stories about the progress, or otherwise, of the war. They contained little that was accurate; how could they? At the time of reporting the journalists were restricted from going on the ground where the action was taking place. All they had to base their stories on was gossip from junior servicemen they encountered on board warships.

Before any journalist was crossed decked to another ship, his name and physical description always preceded him allowing all on board early warning of the arrival of the press. One press story that actually involved me was one filed by Max Hastings, then the *London Evening Standard's* correspondent, but as he was also the most experienced print journalist, his copy was often reproduced in other newspapers. After I had reported, through classified and secure channels, the fact that Argentinian troops were experiencing serious food shortages to the extent that some were suffering malnutrition (see Chapter 13), Max Hastings apparently raised this issue in one of his reports and suggested that there was no truth in this allegation. His contradiction of what could only have come from my reports was based on the fact that as he walked into Stanley on the cessation of hostilities he observed containers full of food supplies. Although I personally never saw them, I do not doubt their existence; however, what I would query was whether Max Hastings was an expert in the supply and demand of food to cater for in excess of 10,000 people for a prolonged period of time. How, for example, was the food to be transported and distributed to those troops not on East Island while serious restrictions were imposed on Argentinian movement to West Island?

Some considerable time after the cessation of hostilities, many reports of the Falkland Islanders' compassion towards the young Argentinian conscripts were revealed. One such Falkland resident, a Mr Goodwin, was quoted as saying the Argentinian soldiers were badly fed; a wagon would come around once or twice a day but all they got was soup. He remembered two young soldiers who would bang on his garden gate waving a container and asking for food and drink. In June 2009, 27 years after the war, the world's press reported that a federal appeals court in Comodora Rivadavia had upheld an earlier decision by a trial court that accusations concerning alleged torture and other crimes were to be considered as possible crimes against humanity, and rejected a petition to abandon proceedings. As a result of this rejection, with over 80 cases under investigation, some 70 Argentinian Army officers were to be charged with murder and causing death by starvation. Argentinian veterans who brought the legal action believe at least four soldiers starved to death while serving on the Falkland Islands. These belated reports are a valuable lesson in not taking all that is written in the British press at face value.

With most of the original British Task Force having returned to the UK and new troops arriving, work had started on repairing the damage caused to the Islands' infrastructure. As this would include extensive work to the runway at Stanley airport, air trooping between Ascension Island and the Falkland Islands was to cease in September until work had been completed. During this period of refurbishment all troop movement would be carried out by ship. Dates for the return of the original crew of the STD had not been given, so on hearing the news of the closure of the airport, pressure was placed on our sponsor branch in the UK to provide reinforcements to allow us to return by air. Almost six months after leaving Portland docks and within days of the Port Stanley airport closing, we were finally on our way back home.

The first part of our journey was by a C130 Hercules aircraft to Ascension Island, where on our arrival we had a quick breakfast, then transferred to a DC 10 and landed at RAF Brize Norton. Of the 3 Brigade Task Force that arrived on the Falkland Islands on 20 May 1982, the crew of the STD were not quite the last troops to leave. On our departure we left behind about fifteen members of the Royal Engineers, part of the original Task Force, who were employed on mine clearance.

16

TRANSITION TO POST-WAR LIFE

When we finally touched down at RAF Brize Norton the majority of the Task Force had returned by sea many weeks before and from what was gleaned from the press, they had been afforded a heroes' welcome with much pomp and ceremony. The passengers on our aircraft accepted that our homecoming would be a little quieter than those previously. What a shock we got when disembarking from the aircraft and we were ushered into the passenger terminal. We had expected at the most to have been met by a few of our wives and families; instead the terminal lounge was full to capacity with people awaiting our arrival. There to welcome us home were representatives from our sponsor branch at the MoD, the CO and others from our parent unit Comms & Sy Gp UK as well as wives, girlfriends and families. All totally unexpected but very much appreciated. At the conclusion of the homecoming reception the crew of the STD, brought together through unusual circumstances and a six-month experience that at times we wouldn't have wished on our worst enemy, felt a sense of personal satisfaction, achievement and professionalism that for most, would never be surpassed during their service career.

Since that late Friday evening in September 1982, with the exception of two or three of the STD crew members, I have had no further contact with all those I met during Operation *Corporate*. One thing that has always remained in my memory was that on saying goodbye to Corporal Jock Cairns, he thanked me for the help and support I had given him and then stated that speaking on behalf of himself and all members of the STD, asked that if circumstances were to be repeated and I found myself in charge of a unit going to war, could they please join me?

After a lengthy period of leave and with the Christmas festivities come and gone, life was back to normal and before I knew it my parent unit was once again preparing to host the senior officers' electronic warfare course. This was held annually, mainly for senior officers in the Royal Corps of Signals about to take command of a regiment, plus other British, American, Australian and Canadian signals officers. I had missed the last course, being on a warship heading south, but now I had no excuse. Nor could I refuse when my OC, Major Fred Searle, asked me to give a presentation on the role of the STD during Operation *Corporate*. However, when he informed me that I had a two-hour slot to fill I wished that I had thought of an excuse.

The date of the course was fast approaching, and because I thought listening to the same person speaking for two hours was too much, I asked WO1 Alan Newman to assist; he would place emphasis on the signals side while I spoke on intelligence aspects. As mentioned at the end of Chapter 5, needing at least one rehearsal to ensure timings and content were correct, a rehearsal was arranged with an audience made up of instructors from the various training wings within the unit. The rehearsal appeared to go quite well, and we finished ten minutes early to allow for any questions. There was complete silence; the last thing a speaker wants when he asks for questions is silence. Looking around our audience I spotted a SNCO who normally was never at a loss for words, and asked him directly if he had any questions. His reply was 'totally gobsmacked'.

When questioned why, he replied that on the return of the Task Force and before I had returned to normal duties, he, along with most of our audience, had been invited to presentations from the Navy, Army, Air Force and Royal Marines, all boasting about how well they had performed during Operation *Corporate* and this was the first presentation he had attended that included the Argentinian perspective. It had changed his thoughts on what had really happened during the war.

Initially, because of what had been said about our presentation, I gave serious consideration to changes in our script but in the end did not make any alterations because what we had said was exactly the way it happened. The presentation proved to be a great success with the senior officers on the course. Through word of mouth it was followed with presentations to selected audiences with the necessary security clearances. I also gave full presentations to staff at the Intelligence Corps Depot in Ashford, Kent, School of Signals, Blandford, Dorset, HQ BAOR, HQ Northern Ireland, HQ 1 BR Corps, British HQ Berlin and GCHQ. A video of one of the presentations even made it to the US, Canada and Australia. This book is a result of being told that I should write one about my experiences at the end of many of these presentations.

Prior to being posted from Comms and Sy Gp UK to a Signals Regiment in BAOR in 1984, because of my involvement in the Falkland Islands War, I was asked to carry out an investigation into the sinking of the Argentinian Cruiser *General Belgrano*, an ex-American ship of Second World War vintage, with the loss of over 360 lives. The then British Prime Minister, Margaret Thatcher, requested a complete and thorough investigation into the sinking as pressure was building from opposition parliamentary backbenchers, in particular Tam Dalyell, for a ministerial enquiry to be held. To carry out my investigation I was provided with every conceivable document, file, report and note imaginable that related to, or included the name, *Belgrano*. Of this large amount of reference material, all of which at the time was covered by the highest security classification, the most useful I found was the Official War Diary from HMS *Conqueror* together with large amounts of strategic intelligence material.

In addition to the British material, I was also given an English translation of a book originally written in Spanish by, I believe, two journalists, one Spanish, the other Argentinian. Their book contained the story of the sinking of the *General Belgrano* from an Argentinian perspective. When I queried why this book had been included amongst the intelligence-related material I had been provided with, I was informed that the book's early publication and release was an attempt by the Argentinians to test the reaction of the British government, hoping they would, when the dust had settled, reverse their comments and allegations made at the time of the sinking and agree with the Argentinians that the warship had in fact left the exclusion zone and was heading back to port in Argentina. Of course, and quite rightly, the British government did not respond openly to anything in this book.

At the time of HMS *Intrepid's* arrival at Ascension, the STD, while not expecting much in the way of intercept of ground forces' radio activity, did expect to intercept the occasional HF communication between the Argentinian Navy and its HQ in Ushuaia in Argentina. On 12 April 1982, the UK announced a maritime exclusion zone within the area 200 nautical miles from the centre of the Falklands, effective at 0400 GMT that day.

However, this exclusion zone did not prevent the *General Belgrano* from entering the zone shortly after setting off from her anchorage at Ushuaia in the early morning of 26 April. While the STD failed to intercept Argentinian Navy transmissions of real intelligence value during the period of commencement of the exclusion zone and the establishing of a bridge-head at San Carlos on 20 May (after this date the Argentine Navy remained in their home ports), the UK collective Sigint organisations produced a substantial amount of material for me to analyse.

It was apparent that the Argentinian Navy might not have taken the introduction of the exclusion zone very seriously, or had possibly taken their 'eye off the ball' when it came to enemy submarine activity, but I was not able to establish why the Argentinian Navy would allow surface vessels to remain within the zone.

From the classified material at my disposal it was evident that HMS *Conqueror*, a nuclear submarine, had been at close quarters with the *General Belgrano* for more than 24 hours before firing three of her straight-running torpedoes, two of which struck their target. The sinking of the *General Belgrano* prompted much condemnation of the British by the Argentine military Junta and the rest of the world, based mainly on Argentine reports that the ship was not in the exclusion zone at the time of the torpedo strike and that she, with her crew and embarked force, was returning to home port.

Hindsight is a wonderful thing and knowing the basic story of the sinking gave me a steer as to what I should look for in separating the wheat from the chaff amongst all the material I had waiting to be analysed. Shortly after the UK's announcement of an exclusion zone, the Argentinian Navy HQ notified its warships, possibly for the purpose of re-grouping, of a pre-arranged rendezvous (RV) point. When the coordinates for this RV were plotted on a map, the actual location, though east of the Falkland Islands, was nevertheless inside the 200 nautical miles exclusion zone. Some considerable time prior to *Conqueror* firing its torpedoes, my analysis revealed that the *General Belgrano* had been instructed to alter course and head in the direction of the RV inside the exclusion zone.

Over the years since the sinking of the *General Belgrano* there has been tremendous controversy as to the direction in which the *General Belgrano* was heading at the time of the attack. The findings of my final report stated that the destination of the vessel was not to her home port as the Argentine Junta stated, but the objective of the ship was to relocate to a prearranged

RV within the exclusion zone. The two significant pieces of intelligence relating to the establishing of the RV and the instruction for the *General Belgrano* to proceed there were both available in real time reporting, but somehow the significance of this intelligence had been overlooked, misread or perhaps had not been read at all by the duty analysts whose responsibility it was to do so. I, on the other hand, was able to analyse the contents of the intelligence material at leisure, not during the 'fog' of war. As I have said, hindsight is a wonderful thing.

The loss of life, in excess of 360 Argentinian personnel, was the greatest single loss throughout the hostilities, greater than the estimated 50 service personnel killed or officially listed as missing with a further 57 personnel wounded, almost all of them badly burned, when *Sir Galahad* was bombed by the Argentine Air Force on 8 June. Many comparisons have been made between the loss of the *General Belgrano* and *Sir Galahad*. My theory why the British suffered fewer deaths and casualties is that once it was obvious *Sir Galahad* was hit, help was on hand from helicopters and landing craft within close proximity, so within a very short period of time they were able to assist in the evacuation of the embarked force. In contrast, while the *General Belgrano* had two accompanying destroyers, both these ships put as much space between them and the stricken vessel as quickly as possible. For some of the survivors who put to sea in their limited number of serviceable lifeboats, it was almost 30 hours before they were recovered.

It is surprising to note that there were several similarities between these incidents – both the *General Belgrano* and *Sir Galahad* were operating in hazardous areas, both vessels had embarked forces lacking in operational experience within a war zone and both forces were inappropriately dressed for the type of operations they were about to carry out at the time of their vessels being hit. Considering the *General Belgrano* was sunk some six weeks before the *Sir Galahad*, no lessons seem to have been learned by the British Task Force during the intervening time.

Quite a long time after the war ended a civil servant was prosecuted under the Official Secrets Act 1911 for the offence of unauthorised disclosure of government information. In February 1985 he was acquitted by a jury acting against the instructions of the judge. I have since wondered if the same civil servant had been privy to my report of the sinking of the *General Belgrano* as requested by the Prime Minister.

Over the years since the end of the war, I have frequently been asked the same two questions. Firstly, was it right for the British to enter a war over the possession of a few small islands located miles from British soil in an extremely inhospitable area of the South Atlantic? My answer then, as it is today, without getting into the politics of the question, is a definite yes. Furthermore, not one person of the British Task Force with whom I spoke throughout the entire period of hostilities and for some time after, failed to agree with me that the action taken by Thatcher and her Cabinet was fully justified and the only acceptable thing to have done.

The second question is whether I have ever had the misfortune to suffer any form of 'traumatic distress', either during Operation *Corporate* or since? This I find a difficult question to answer. On my return from the Falkland Islands in September 1982, I was not aware that I was in any way differ-ent from the person that had left some six months earlier. I was extremely pleased to have returned with a full compliment of the detachment with whom I departed, and while I had often complained bitterly about the conditions and not appreciated fully the thinking behind some of the things I had to endure, in the end it all turned out fine.

Yet, according to my wife, family and those close to me, without my realising it I had changed as a result of being in a war zone. It was difficult to state exactly how I was different but there were changes were to my personality and the way I responded to and treated those closest to me. Perhaps the reason for the changes not being even more pronounced was that apparently I never stopped talking for about three months after my return. This excessive talking may have prevented any trauma from taking hold, even though the minor changes to my personality have remained to this day. In contrast, one of the STD crew whom I met socially several years after the end of Operation *Corporate* said that going to war had severely affected him. For several years he had had sleepless nights or had nightmares about the Falklands. He had not spoken to anyone about his experiences.

The Falklands War, and indeed my entire Army career, affected me in one other serious way. From a very young age it was necessary for me to wear headphones through which I listened to sound waves of various levels; barely perceptible at around 3dB, to dangerous levels of between 115 and 140dB. The result of this is that the hearing in one of my ears has been degraded to just about an acceptable level, while in the other I am profoundly deaf to the extent that even the most advanced technological hearing aids are of no use. Isn't it ironic that after so many years of being a Silent Listener, I will now forever listen to a kind of silence?

Argentinian Marines surrender at Government House.

APPENDIX 1

'LA GACETA ARGENTINA'

Port Stanley, 26 May 1982
Year 1 Number 7 Special Edition

The part played in the war by the Guemes Combat Team (C Company 25th Infantry Regt) as forwarded to the Commander of 1 Brigade of III Corps at the end of the harassing operations carried out against the enemy who landed in the San Carlos area.

1   It was at 0230 hours when, from the Command Post of 'Aguila' (Officer Commanding the Guemes Combat Team, Port San Carlos) we heard incoming naval gunfire landing in an area close to hill 324.

2   Aguila radioed to 'Gato' (Commander of the Heavy Weapons Section located on hill 324), Gato did not reply to any of our calls until 0600 hours.

3   The Naval bombardment was intermittent, lasting from 180 minutes, on an undetermined area.

4   Aguila was awaiting a messenger from Gato as they were unable to establish radio contact.

5   At 0630 hours Aguila placed observers with night vision capability on the heights of Port San Carlos.

6   It was at 0810, first light, that one observer made out a huge white vessel (not a warship) at the entrance to Port San Carlos water.

7   At 0810 Aguila moved to the height and with the aid of night vision equipment was able to discern behind the white vessel at least three frigates.

8   At 0820 Aguila saw a large landing craft (LCM) much bigger than normal landing craft (LSU), pull away from the white vessel and head for San Carlos settlement, various helicopters were over flying the vessels.

9   At 0822 they were able to make out landing craft heading in all directions.

10   At 0830 the higher observation posts of Aguila reported that British infantry were advancing in files from the west.

11   At 0831 Aguila informed his command that they would defend their position.

12   Aguila ordered his troops to the eastern heights of the port in order to escape the encirclement that the British infantry were trying to achieve.

13   At about 0840 dozens of British infantry fell on the vacated Port San Carlos, and at the same time a Sea King helicopter arrived from the east to complete the encirclement.

14   The order to open fire on the enemy helicopter was given; it was very badly damaged and decided not to land in the port area. It escaped from the area.

15   The British infantry opened fire with out reaching Aguila's position.

16   One minute later, a Sea Lynx helicopter approached Aguila's position in order to fire its rockets. Fire was opened up on it with all weapons and it fell into San Carlos water sinking immediately, a body was floating on the surface, the other (crewman) swam to a buoy, a launch rushed to his aid.

17   Shooting down the helicopter had marked our position and the infantry opened up with mortar fire, failing to hit the target.

18   Aguila ordered another change of position, more to the east, in order to escape the mortar fire.

19   A Sea Lynx helicopter appeared over the new position firing a machine gun and trying to get into position to fire its rockets. Again the order was given to open fire and the helicopter fell to the ground in flames, it came down about 10 metres from Aguila's position and it was possible to check that the three crew members were dead.

20   The mortars reopened fire but could not locate us accurately.

21   Aguila ordered yet another change of position and three minutes later the enemy sent yet another Sea Lynx apparently to direct naval gunfire.

22   Fire was again opened upon the machine and the pilot managed to withdraw with his helicopter smoking and severely damaged.

23   Naval and mortar fire was opened up but it was landing about 500 metres away, they could not pinpoint our position.

24   During the 20-25 minutes that the fight with the helicopters lasted, there were about 200 British infantry in Port San Carlos settlement and from the way the landing craft were heading for San Carlos settlement they must have landed double that number there.

25   Aguila had only 1 x rifle section and the headquarters and logistic platoon of the company.

26   Aguila ordered a move to new positions.

27   At about 0930 he was able to observe, from these positions, a furious attack on the British vessels by our aircraft.

28   At the same time the vessels stopped their heavy gunfire on Aguila's position and tried to counter the air attack.

29   At no time did the British infantry try to close in on Aguila's position, the effect of their small arms fire was virtually nil and although their mortars maintained a heavy fire they did not hit their target.

30   For some three hours the troops of Aguila awaited the re-deployment of Gato from Hill 234.

31   During the fighting the troops of Aguila suffered no losses except for heavy personal kit, which was left in Port San Carlos and an (Instalaza) Rocket Launcher which was unserviceable after being hit by a machine gun round from the British helicopter which had been shot down.

32   The damage caused to the enemy was follows:

2 Marines killed
2 Sea Lynx helicopters shot down (of the two crews only 1 man survived)
A Sea King badly damaged
A Sea Lynx damaged and certainly unserviceable.

33   The troops of Aguila found themselves in the position of the bearers of bad news when they informed Capanga of the failure of the combat team. 2 x officers, 9 x NCOs and 31 men.

34   With Gato who had not made contact were 1 x Officer, 4 x NCOs and 15 men.

35   During the operations, Aguila was able to discover that the enemy inserted Reconnaissance patrols by Sea King helicopter at very low level. These reconnaissance ops were totally defensive but on encountering the firepower of Aguila during the night ceased to advance and withdrew without getting into action.

36   During the fight the population of San Carlos mocked the Argentinian troops, insulting them and making gestures, when the helicopters were downed they rushed to the aid of the crews. This proves that the people are not hostile, because they are afraid. But they change quickly when they have might on their side. They guided the British to our position with signs.

37   As these actions were going on, it was noted that the enemy were slow at aiming and lacking in firepower, particularly the helicopter crews who allowed the infantry plenty of time to shoot them down easily.

Carlos Daniel Esteban
Senior Lt
OC C Company 25th Infantry Regiment

APPENDIX 2

COUNTERATTACK NOTES

Copy of the original handwritten notes made by the Records Officer on a meeting he attended to discuss the counterattack by the Argentine Ground Forces on West Island.

- FT Recon atacará p/ conq Darwin y continuar atq
en direccion N y NO a fin de contribuir al cumpli-
miento de la mision de la Cr I II.

Tareas implicitas
- Confeccion plan embarque y trasporte.
- Ejecutar marcha táctica.
- " atq.
- Conq y consolid obj.
- Continuación atq a orden.

Tarea explicita:
- Preparar Unidad.
- " plan alist de 24 hs.
- Ejec marcha hasta Topo.

- Actualizar necesidades reemplazos de personal en
rubro del elon asalto.

b. Fases.
Fase agua: (Secundaria) Maritima y desembarco, desde PC 4
hasta PC 5 y PC 6 (Tentativo)
(secundaria) Aeromovil, desde PC 4 hasta PC 7

Fase 2 Rodeo Tierra. constitución de cabeza aérea para
rodeo de DA y A y base operaciones logisticas. Marcha
Táctica inmediata por escalon asalto, camino tentativo
PC 5, PC 14, PC 2.
Diurno y o nocturno (elon ligero y maximo poder
combate)
Elon retag: una vez asegurado el frente.
PT marcha: PC 9. 8horas despues desembarco.
Fase Tucuman: ataque y conq obj, 2 horas despues
de terminada la marcha
Fase Rodeo Salta: Defensa (prevision contrataque) desde PC 3).
Existencia campo minado y trampas explosivas, control de
la poblacion. Restauracion de pistas.
Fase 5 Tanque anfibio maipu: continuacion ataque.
Variante 1: PC 2 y/o PC 5

Variante 2: L C D 3 , PC 16 , PC 17.

Variante 3: ambas.

c. F T "Reconquista"

Misiones part: transporte, marcha, ataque, defensa, previsiones
continuación de ataque,

d. Apoyo: prever aproximación fgo naval y aeronaves, oportuni-
dad, fase tucumán

x. Instrucciones coordinación

Dea "D" a determinar.

Preaviso: 24 horas.

The 'Y' services plaque at the National Memorial Arboretum.

INDEX